Dose Optimization in Digital Radiography and Computed Tomography

Euclid Seeram

Dose Optimization in Digital Radiography and Computed Tomography

An Essential Guide

Springer

Euclid Seeram
University of Canberra
Burnaby, BC, Canada

ISBN 978-3-031-22870-4 ISBN 978-3-031-22871-1 (eBook)
https://doi.org/10.1007/978-3-031-22871-1

This Springer imprint is published by the registered company Springer Nature Switzerland AG
The registered company address is: Gewerbestrasse 11, 6330 Cham, Switzerland

This book is dedicated to my two smart, precious, and overall cute granddaughters, Claire and Charlotte, with love and blessings to you both forever.

Preface

Digital radiography (DR) and computed tomography (CT) are based on physics, engineering, and computer science principles that have influenced their survival as useful clinical tools for imaging the patient. These two modalities have experienced several technical innovations in recent years. These innovations have played major roles in not only reduction of the radiation dose to the patient but improving the image quality needed for diagnostic interpretation. For example, the fundamental concepts of the wide exposure latitude of DR systems and the standardized exposure indicator established by the International Electrotechnical Commission (IEC) have provided researchers with the motivation to operate DR systems with the goal of optimization of the dose and image quality. Additionally, CT technical innovations such as, for example, new detector technology, iterative reconstruction (IR) algorithms, and artificial intelligence-based image reconstruction are now offered by several CT vendors, and play a significant role in dose reduction and optimization in CT. Furthermore, the introduction of photon counting detectors (PCDs) has solved the major problem of image noise during low-dose CT imaging. Dose optimization has now become an integral part of DR and CT practice, especially in radiation protection of the patient.

This book, *Dose Optimization in Digital Radiography and Computed Tomography*, provides yet another useful resource to meet the educational requirements of professional radiologic technology associations including the *American Society of Radiologic Technologists* (ASRT), the *American Registry of Radiologic Technologists* (ARRT), the *Canadian Association of Medical Radiation Technologists* (CAMRT), the *College of Radiographers* in the United Kingdom, as well as those professional medical imaging organizations in Africa, Asia, Australia, and continental Europe. Additionally, this book may serve as a resource for biomedical engineering technology programs that include DR and CT imaging systems in their curriculum; residents in radiology, and medical physics students studying the use of DR and CT in medical imaging.

There are seven chapters in this book as follows:

Chapter 1 provides the *motivation for dose optimization in DR and CT, with current and relevant references.*

Chapter 2 deals with a *technical review of DR technologies, including computed radiography (CR) and flat-panel digital radiography (FPDR), with current and relevant references.*

Chapter 3 identifies and describes the *major technical factors affecting the dose in DR,* and various *optimization strategies* in DR, *with current and relevant references.*

Chapter 4 addresses the *technical aspects of CT, with current and relevant references.*

Chapter 5 takes a close examination of the *technical factors affecting the dose in CT, dose reduction, and dose optimization strategies, with current and relevant references.*

Chapter 6 explains the nature of *dose optimization research in medical imaging* and presents an *overview of observer performance methods in image quality assessment, with current and relevant references.*

Chapter 7 is the final chapter and provides a *set of self-assessment multiple-choice questions* (with answers) to check your understanding of the materials studied.

Enjoy the content and questions that follow and best wishes for any examination that you have to write and pass. Remember, your patients will benefit from your wisdom.

Burnaby, BC, Canada Euclid Seeram

Acknowledgments

One of the most satisfying tasks in writing a book of this nature is to express gratitude to all those who have done the original work. These are the medical physicists, biomedical engineers, computer scientists, radiologists, and technologists who are working in both research and practice to conceptualize and invent better ways to image the patient with reduced dose and improved image quality.

This book deals with two major imaging modalities, *digital radiography* (DR) and *computed tomography* (CT), and in this respect, it is indeed a pleasure to express sincere thanks to all scientific and clinical experts in the field of DR and CT.

The content of this book is centered around the published works and expertise of several noted medical physicists, radiologists and imaging technologists, computer scientists, and biomedical engineers, whose research on dose optimization have been quoted here to support the use of the ALARA (as low as reasonably achievable) principle in radiation protection of the patient, established by the International Commission on Radiological Protection (ICRP).

I am most grateful to Dr Rob Davidson, PhD, MAppSc (MI), BBus, FASMIRT, Professor of Medical Imaging, University of Canberra, Australia. Dr. Davidson has taught DR and CT physics and instrumentation for decades. He was one of my PhD supervisors who guided me through the experiments on dose optimization in DR imaging. He has also been responsible for my adjunct professorship appointment at the University of Canberra, and has supported all my other projects linked to DR and CT physics and technology. Thanks mate.

Furthermore, two medical physicists to whom I am truly grateful for my DR education are Dr. Anthony Siebert, PhD, of the University of California at Davis and Dr. Charles Willis, PhD, formerly of the University of Texas, MD Anderson Cancer Center, from whom I have learned the physics and technical aspects of digital radiography through their seminars and workshops that I have attended.

My CT physics education stems from not only attending conferences and workshops but also personal communications with a number of notable medical physicists. In particular, I owe a good deal of thanks to Dr. Godfrey Hounsfield and Dr. Allan Cormack, both of whom shared the Nobel Prize for Medicine and Physiology for their work in the invention and development of the CT scanner. I have been in personal communication with Dr. Hounsfield, and he graciously provided me with details of his biography and his original experiments, and signed his Nobel Lecture, which he sent to me. Other CT experts from whose published works I have learned a great deal and to whom I am indebted include Jiang Hsieh, PhD, former Chief Scientist with General Electric Healthcare; world-expert CT physicist, Professor Willi Kalender, PhD, Institute of Medical Physics in Germany; Mahadevappa Mahesh, PhD, Chief Medical Physicist, Johns Hopkins Hospital in Baltimore; Michael McNitt-Gray, PhD, University of California; Cynthia McCollough, PhD, Mayo Clinic; and Thomas Flohr, PhD, Siemens Healtineers, Germany.

Another notable individual who has made an impact on my professional journey is Valentina Al Hamouche, MRT, MSc, and CEO for Vision, Compassion, Awareness (VCA) Education Solutions for Health Professionals Inc. Toronto, Ontario, Canada. Valentina has given me the opportunity to be a regular guest lecturer in not only face-to-face presentations but also live webinars on CT physics and instrumentation, and radiographic sciences as well as live webinars on dose optimization in CT and DR to technologists all over the world. She has brought continuing education to health professionals through her Vision, Compassion, and Awareness, the name of her educational organization (www.VCAeducation.ca)

I am particularly grateful to Merry Stuber, Senior Editor, Cell Biology & Biomedical Engineering at Springer, a part of Springer Nature, New York, NY, USA, who did all the hard work in reviewing the proposal herself but also in getting external reviews of the proposal that led her to accept it for publication. Merry has provide the needed continuous support and encouragement to bring this work to fruition. Additionally, I am grateful to members of the production team at Springer Nature, who have worked exceptionally hard during the production of this book, especially in the page-proof stages.

I humbly acknowledge the support and praise that I get from my beautiful family. First my lovely wife, Trish, a warm, smart, caring, and a very special person in my life, thanks babe. Secondly, my caring and very brilliant son David, the best dad on the planet to his two most precious daughters, *my granddaughters*, to whom this book is dedicated. Thanks for your enduring love, support, and encouragement.

Last but not least, I must thank my students, not only in Canada but also all over the world, who have diligently completed my DR and CT physics and technology courses, both at the diploma and degree levels. Thanks for all the challenging questions, which have always "kept me on my toes."

Keep on learning and have fun answering the questions that follow, and remember, your patients will benefit from your wisdom.

Burnaby, BC, Canada Euclid Seeram

Contents

1 Dose Optimization: A Major Principle of Optimization 1
1.1 Biological Effects of Radiation Exposure 1
1.2 Fundamental Principles of Radiation Protection 2
1.2.1 The Principle of Justification 3
1.2.2 The Principle of Optimization 3
1.3 Motivation for Dose Optimization in Digital Radiography and Computed Tomography 6
1.3.1 Radiation Risks 6
1.3.2 Dose Creep in Digital Radiography 6
1.3.3 Dose in Computed Tomography 8
1.4 Optimization Strategies: An Overview 9
References 9

2 Digital Radiography: A Technical Review 13
2.1 Digital Radiography: Two Notable Facts Related to Dose Management 13
2.2 Digital Radiography Systems: A Technical Review 15
2.2.1 Computed Radiography: Physical Principles and Technology 16
2.2.2 Flat-Panel Digital Radiography: Physical Principles and Technology 17
2.3 Other Considerations Related to CR and FPDR Systems 19
2.3.1 Image Processing 19
2.3.2 The Standardized Exposure Indicator and the Deviation Index 20
2.4 Factors Affecting the Dose in Digital Radiography: An Essential Review 22
2.5 Diagnostic Reference Levels 22
2.5.1 Definitions and Major Guidelines 23
References 23

3 Optimization Strategies in Digital Radiography . . . 25
3.1 Introduction . . . 25
3.2 Optimization of Exposure Technique Factors . . . 26
3.2.1 Optimization of kV and mAs . . . 26
3.3 Optimization of the Exposure Indicator . . . 27
3.4 Optimization of the Deviation Index . . . 30
3.5 Optimization Using Image Postprocessing Algorithms . . . 31
3.5.1 Optimization Studies Using Multifrequency Processing and Noise Reduction Algorithms . . . 32
3.6 Diagnostic Reference Levels: A Useful Tool in Dose Optimization . . . 36
References . . . 37

4 Computed Tomography: A Technical Review . . . 41
4.1 Introduction . . . 41
4.2 The CT Scanner: Fundamental Physics and Major System Components . . . 42
4.2.1 Attenuation Physics: An Essential Overview . . . 42
4.2.2 Attenuation and CT Numbers . . . 43
4.2.3 Data Acquisition: System Components and Principles . . . 44
4.2.4 Image Reconstruction . . . 46
4.2.5 Image Display/Storage/Communication . . . 49
4.3 Multi-slice CT Principles at a Glance . . . 50
4.3.1 MSCT Detectors: An Overview . . . 51
4.3.2 Selectable Scan Parameters . . . 53
4.4 Radiation Protection . . . 54
References . . . 55

5 Dose Reduction and Optimization Strategies in Computed Tomography . . . 57
5.1 Introduction . . . 57
5.2 Motivation for Dose Optimization in CT . . . 58
5.2.1 Literature on Cancer Risks from CT . . . 59
5.2.2 Effects of Low-Dose Chest CT on Chromosomal DNA . . . 60
5.2.3 In Summary . . . 60
5.3 Radiation Protection Principles . . . 61
5.4 Elements of CT Dosimetry at a Glance . . . 61
5.5 Factors Affecting the Dose in CT: An Overview . . . 63
5.5.1 Image Quality Considerations . . . 63
5.5.2 Significant Dose Parameters . . . 63
5.6 Dose Optimization Strategies in CT . . . 65
5.7 Other Useful Tools for Dose Reduction and Optimization in CT . . . 66
5.7.1 The Use of DRLs . . . 66
5.7.2 The Use of AI-Based Image Reconstruction . . . 68
5.7.3 The Use of PCDs . . . 70

5.8 The Role of the Technologist in CT Dose Reduction and Optimization 71
References 72

6 Optimization Research in Medical Imaging 77
6.1 Introduction 77
6.1.1 Strategies Used in Dose/Image Quality Optimization Research 78
6.1.2 The Nature of Image Quality 79
6.1.3 Image Quality Assessment Tools for Clinical CT Images: An Overview 80
6.1.4 Receiver Operating Characteristics: A Brief Overview 81
6.1.5 Visual Grading Analysis: An Overview 82
6.1.6 Example of a Dose Optimization Study in Digital Radiography 84
References 85

7 Review Questions 87

Index 101

Chapter 1
Dose Optimization: A Major Principle of Optimization

Keywords Biological effects of radiation exposure: International Commission on Radiological Protection (ICRP) · Principles of radiation protection · Justification · Optimization · Dose optimization · Dose creep in digital radiography · Doses in computed tomography (CT) · Optimization strategies

1.1 Biological Effects of Radiation Exposure

Biological effects of radiation exposure are sometimes referred to as health effects, harmful effects, or simply radiation risks. These effects have been studied extensively and documented in what is popularly known as *epidemiologic studies* [1]. These studies have identified several sources which have been placed into the following classes: Early radiation workers such as radiologists and physicists who were exposed to high doses; workers from radiation and nuclear industries; survivors from Hiroshima and Nagasaki (H-N) atomic bomb explosions; and workers and inhabitants exposed to nuclear reactor accidents such as those at Chernobyl and Three Mile Island. Other significant resources on biological effects are available from the Biological Effects of Ionizing Radiation (BEIR) Reports, which also includes data on the exposure of patients to high doses in medical imaging. Furthermore, data on the H-N survivors have been provided the Radiation Effects Research Foundation (RERF).

There are at least two significant and important points on biological effects has been identified and stated by Hendee and O'Connor [1]. The first point is as follows:

> The RERF studies of the Japanese atomic bomb survivors are the major source of what is known about the health consequences to individuals exposed to ionizing radiation. The RERF data and the models of radiation injury developed by the RERF scientists form the back bone of the BEIR VII report. It is from the summary tables of radiation risk in the BEIR VII report that projections of cancer incidence and death are made for medical exposures in the United States. Hence an analysis of the assumptions and limitations of risk estimates derived from BEIR VII must include a review of the RERF studies from which BEIR VII is derived. (p. 314)

E. Seeram, *Dose Optimization in Digital Radiography and Computed Tomography*,
https://doi.org/10.1007/978-3-031-22871-1_1

The biological effects identified in the above-mentioned reports have been placed into two categories, namely; stochastic effects and deterministic effects. *Stochastic effects* are those effects for which the probability of occurrence in the individual exposed, increases with increasing dose and for which there is no *threshold dose*. Examples of stochastic effects include cancer, leukemia, and hereditary effects. *Deterministic effects* on the other hand, are those effects for which the severity of the effect in the exposed individual increases with increasing dose, and for which there is a threshold dose. Examples of deterministic effects include cataracts, radiation-induced skin burns, tissue damage, and organ disfunction [2].

The second significant and important point is that:

> RERF data provide statistically significant evidence of increased cancers in Japanese survivors who received doses of 100 mSv and higher with cancer incidence appearing to increase linearly with dose. At less than 100 mSv, an increase in radiation-induced cancers, if any is too small to be distinguishable from cancer incidence due to all causes. Consequently, a model must be deployed to extrapolate from radiation-induced cancers at doses greater than 100 mSv to a hypothetical and much smaller number of cancers induced by doses of a few millisieverts delivered during medical imaging [1].

Subsequently, several *dose-response models* for *extrapolating from a high dose situation* (for example 100 mSv) *to the low doses used in diagnostic x-ray imaging* have been proposed, however, only two will be briefly outlined in this chapter, that is; the *linear dose-response model without a threshold dose*, known as the *LNT model*; and the linear dose-response model with a threshold dose. While the former implies that there is no safe amount of radiation dose, and that any dose no matter how small carries a degree of risk; the latter implies the effect of radiation exposure will manifest itself at a certain dose referred to as the threshold dose. Below this threshold no effect is observed [2].

Currently, there is ongoing debate about these models and their use in medical imaging, however, Hendee and O'Connor [1] state that "the model used most widely is the LNT model. This model is not chosen because there is solid biologic or epidemiologic data supporting its use. Rather, it is used because of its simplicity and because it is a conservative approach (i.e., if it is not correct, then it probably overestimates the risk of cancer induction at low dose). For the purpose of establishing radiation protection standards for occupationally exposed individuals and members of the public, a conservative model that overestimates the risk is preferred over a model that underestimates risk." (p. 316).

1.2 Fundamental Principles of Radiation Protection

In order to protect patients undergoing medical imaging procedures using ionizing radiation, the ICRP has recommended a comprehensive *framework of radiation protection* against ionizing radiation. This framework is intended to prevent deterministic effects from occurring by keeping the radiation doses below threshold values; and to minimize the probability of stochastic effects by taking all reasonable efforts.

The framework is a recommendation that addresses three fundamental principles of radiation protection [3]; *justification*, *optimization*, and *dose limitation*. While justification and optimization focus on exposed individuals, the third principle examines both occupational and public exposures, and does not include medical exposures. These principles are now accepted by various national radiation protection organizations such as for example; the National Council on Radiation Protection and Measurements (NCRP) in the United States; Radiation Protection Bureau-Health Canada; the National Radiological Protection Board (NRPB) in the United Kingdom; and other radiation protection organizations around the world.

This textbook will only address the major elements of optimization.

1.2.1 The Principle of Justification

The principle of *justification* is described as follows by the ICRP:

> Any decision that alters the radiation exposure situation should do more good than harm. This means that, by introducing a new radiation source, by reducing existing exposure, or by reducing the risk of potential exposure, one should achieve sufficient individual or societal benefit to offset the detriment it causes [3].

Justification therefore focusses on the concept of *net benefit*, that is, there must be a benefit related to the introduction of any new imaging modality, procedure, or radiation exposure. The responsibility for justification falls upon not only the physician who requests x-ray examinations based on good clinical reasons, and the individual who performs the examination [4].

1.2.2 The Principle of Optimization

Following justification, the principle of optimization is applied. The ICRP states that the essential goal of optimization during the conduct of the examination is "to adjust imaging parameters and institute protective measures in such a way that the required image is obtained with the lowest possible radiation dose, and net benefit is maximised" [3]. In this regard, the ICRP recommends that the ALARA (as low as reasonably achievable) philosophy should be applied. Furthermore, the ICRP [3] emphasizes that ALARA "is only part of the concept of optimisation. The entire concept implies, more precisely, keeping patient exposure to the minimum necessary to achieve the required medical objective (diagnostic or therapeutic). In diagnostic imaging and x-ray-guided interventions, it means that the number and quality of images are sufficient to obtain the information needed for diagnosis or intervention…."

Optimization involves at least four elements [5], namely; the imaging equipment, the adequacy of the imaging equipment, technical parameters of the examination, and diagnostic reference levels, and requires that:

- The *imaging equipment* is in good working order to deliver correct exposures for respective examinations; and complies with required performance standards established by various regulatory and professional standards. Such equipment is subject to acceptance testing when it is installed; and subsequently, routine quality control tests should be conducted during routine use.
- The *adequacy of the equipment* takes into consideration not only the design and use of a wide range of settings, but also specific pre-installed protocols and various approaches for dose reduction when imaging pediatric patients.
- *Technical parameters for the examination* refer to several dose reduction methods, such as the use of correct exposure technique factors (mAs and kV) for the examination, collimation, filtration, removable grids, and pulsed fluoroscopy, to mention but a few.
- *Diagnostic reference levels* (DRLs) be implemented in the assessment of the medical exposure of patients. There are several definitions of the DRL, however, only two will be provided in here. The DRLs as defined by the ICRP "are a form of investigation level applied to an easily measured quantity, usually absorbed dose in all, or tissue-equivalent material at the surface of a simple standard phantom or a representative patient" [6].

The third principle of radiation protection is dose limitation. According to the ICRP this principle deals with the application of dose limits, and requires that "the total dose to any individual from regulated sources in planned exposure situations other than medical exposure of patients should not exceed the appropriate limits recommended by the Commission" [3]. Since this principle specifically addresses doses to occupationally-exposed individuals (technologists for example), members of the public, and doses to other body parts such as the skin, lens of the eyes, and extremities, it will not be discussed further in this book.

Another characteristic feature of optimization is that the ALARA philosophy includes image quality, that is the required image quality should be obtained with the lowest possible radiation dose. Therefore, dose optimization strategies and research must consider image quality and dose.

Digital Radiography and Computed Tomography: System Components at a Glance

Dose optimization in digital radiography and computed tomography (CT) requires a thorough understanding of not only the major technical components but also the factors affecting the dose to the patient, in each of these two imaging systems. Such understanding provides the basis for how the dose from each of these imaging technologies can be optimized.

While digital radiography (DR) essentially includes two major and popular systems, commonly referred to as computed radiography (CR) and flat-panel digital radiography (FPDR) [2, 6], the technical components of which are illustrated in Fig. 1.1; CT systems are more complex, not only in terms of physical principles, but also the technologies involved in creating the CT image [7]. Major CT system components are illustrated in Fig. 1.2. The principles and technology of DR and CT systems will be described in Chaps. 2 and 4 respectively.

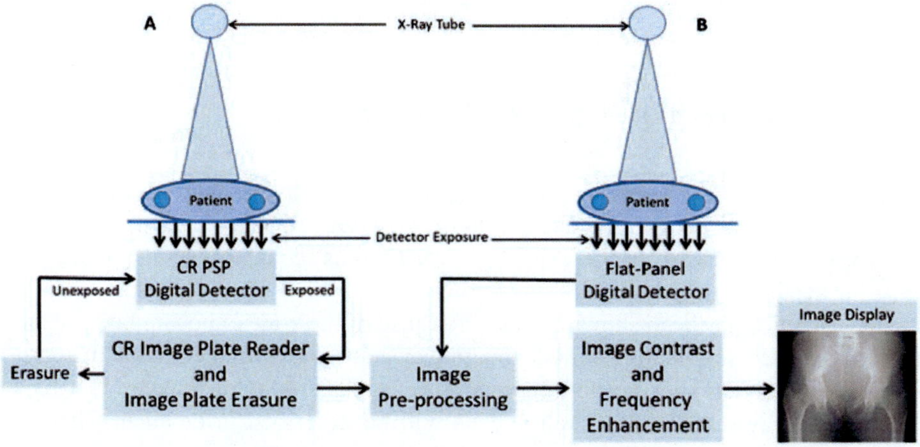

Fig. 1.1 The major technical components of two commonplace digital radiography imaging systems; (**a**) computed radiography (CR) and (**b**) flat-panel digital radiography (FPDR). (See text for further explanation)

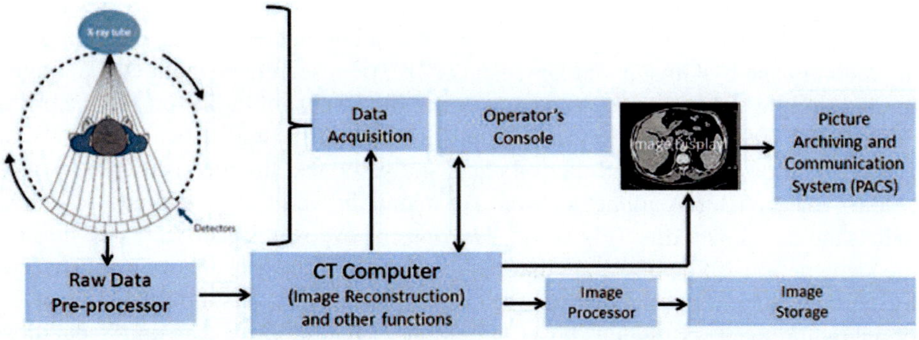

Fig. 1.2 The primary technical components of a CT Imaging system, includes data acquisition, image reconstruction, and image display, storage, and communications. (See text for further explanation)

1.3 Motivation for Dose Optimization in Digital Radiography and Computed Tomography

1.3.1 Radiation Risks

Radiation risks provide the first and most important motivation for dose optimization in diagnostic imaging using x-rays. These risks are real and have been described extensively in the literature [8, 9] and fall into two categories, namely; *stochastic effects* and *deterministic effects*. These risks have been summarized in 1 above.

1.3.2 Dose Creep in Digital Radiography

The motivation for dose optimization in digital radiography has its roots in what has been referred to as *dose creep* or *exposure creep,* which has been identified in the literature as an increase in the exposure over time as technologists use these DR imaging systems [10–14].

This practice has been shown to occur because of the wide exposure latitude and the image processing features of DR imaging systems. Such image processing compensates for underexposure and overexposure while producing images that have acceptable gray scale image quality. This will be described in detail in Chap. 2. While underexposed images appear noisy (quantum mottle), overexposed images do not exhibit quantum noise [15]. Noise-free images are always preferred by radiologists. The significant and important problem with the use of overexposure is an increase in the dose to the patient. Therefore, it is mandatory that users operate with the ALARA philosophy in mind. This provides the opportunity for dose optimization in digital radiography.

An important and meaningful study demonstrating that dose creep is a real phenomenon is one by Gibson and Davidson [12]. The researchers conducted a longitudinal study which examined and categorized the Exposure Indicators (EIs) s from more than 17,000 mobile chest radiographs from intensive and critical care units at one hospital over a 26-month period. This study showed that dose creep occurred when x-ray exposure parameters were set manually, and that overexposure index values increased over time (Fig. 1.3a). The optimal exposures also decreased during the same period. To demonstrate that the increase in overexposures was not caused by external factors, an intervention was needed. The researchers introduced an additional requirement for mobile chest radiographs in the intensive care unit; technologists were to record exposure factors of kilovolt (kV), milliampere-seconds (mAs), and source-to-image receptor distance, as well as the EI value for each examination. After the intervention, the initially observed trend of overexposures was gradually halted over a 14-month period. Figure 1.3b illustrates the halt of the overexposure

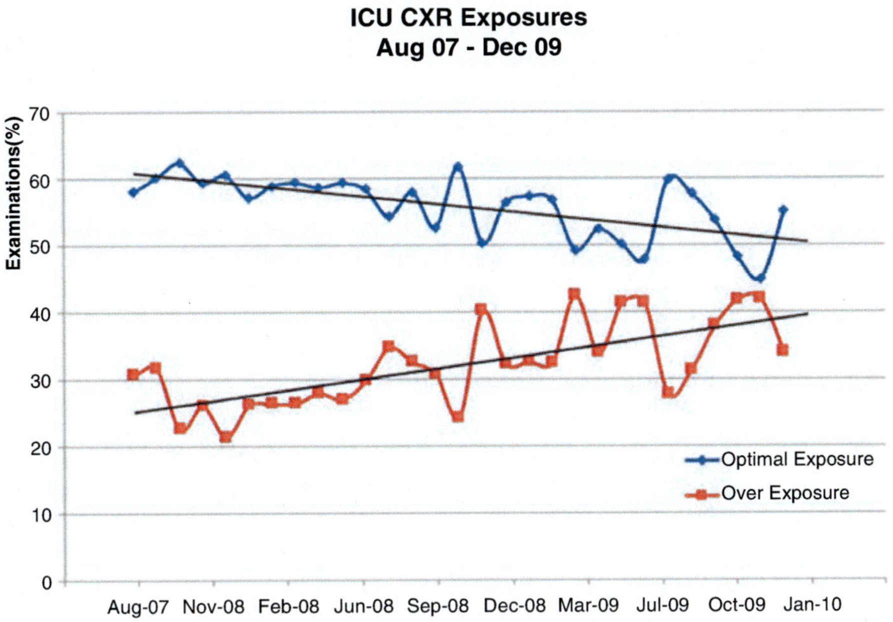

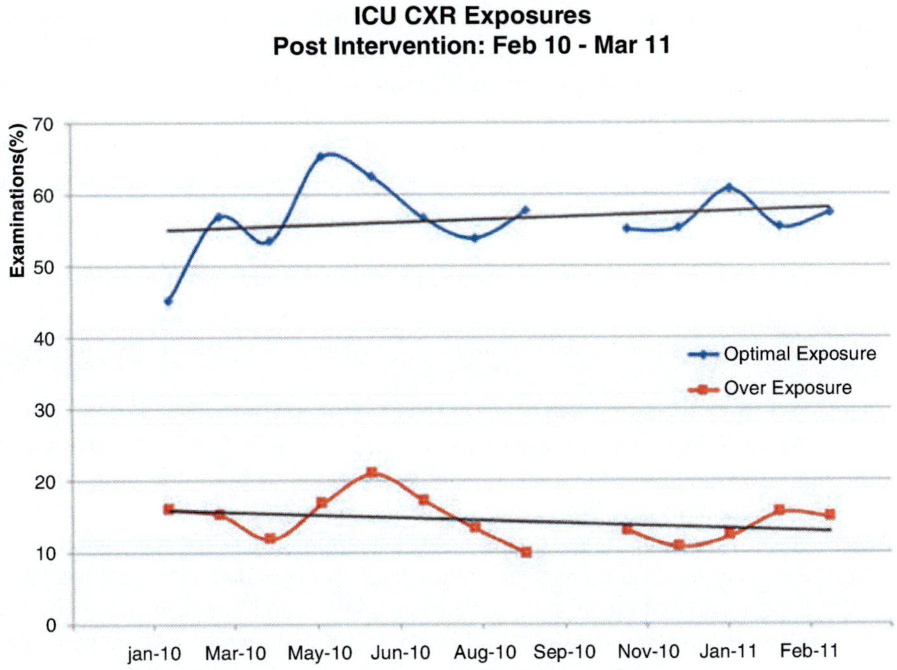

Fig. 1.3 Graphs comparing the exposure index values for mobile chest radiographs performed in the intensive care unit from before and after intervention. Before intervention, overexposure index values increased over time, indicating dose creep in digital radiography (**a**). Optimal exposures decreased over time, stopping only after an intervention was applied (**b**). (Reproduced with permission from Professor Rob Davidson, Medical Imaging; School of Health Sciences, Faculty of Health, University of Canberra, Australia)

trend. This intervention demonstrated conclusively that dose creep is a challenge that can be addressed at the systemic level.

In another study of dose creep, Dalah [16], quantified dose creep for skull and chest examinations using dose area product and entrance surface dose. The results showed that the chest examination had the highest estimated dose creep of 51%. Dalah [16] concluded that "dose-creep can be considerably high raising the need for thorough investigations and society awareness". This is yet another dose study that supports the need for dose optimization in digital radiography.

1.3.3 Dose in Computed Tomography

The first clinically useful CT scanner was invented by Godfrey Newbold Hounsfield, who worked at Electric and Musical Industries (EMI) in the United Kingdom. Another notable pioneer in the is the evolution of the CT scanner is Allan MacLeod Cormack, in South Africa. Hounsfield and Cormack shared the 1979 Nobel Prize in Medicine or Physiology for their significant contributions such as providing mathematical solutions to the problem in CT [7].

The use of CT in in clinical medicine has increased significantly ever since its introduction in the 1970s, due to several technical advances in CT scanner design and performance. Several studies have shown that patient doses from CT examinations are high relative to other radiography examinations [17–19] and that *CT doses typically range from 5 to 50 mGy* to each organ within the image field [20]. As of 2011, CT contributed the highest collective amount of medical radiation exposure in the United States compared with any other medical imaging modality [21]. These relatively high doses from CT examinations, especially in pediatric imaging, have raised concerns related to the risks of cancer associated with CT imaging [1, 2, 17–22].

In addition, continued technological advances in CT, such as for example, more sensitive detectors, iterative and deep learning reconstruction algorithms, have played a key role in reducing the dose from CT scanners, for specific use in what is popularly referred to as *low-dose CT imaging.*

The Food and Drug Administration (FDA) [23] in the United States, in an article entitled "What are the Radiation Risks from CT" states that "the effective doses from diagnostic CT procedures are typically estimated to be in the range of 1 to 10 mSv". Additionally, the World Nuclear Association [24] in 2022, states that the effective dose from an abdomen and pelvis CT examination is 10 mSv.

More recently, Smith-Bindman and collaborators [25] conducted an extensive prospective cohort study, "to determine patient, institution, and machine characteristics that contribute to variation in radiation doses used for computed tomography (CT)". The researchers found that "CT protocols and radiation doses vary greatly across countries and are primarily attributable to local choices regarding technical parameters, rather than patient, institution, or machine characteristics. These findings suggest that the optimization of doses to a consistent standard should be possible".

1.4 Optimization Strategies: An Overview

In addressing optimization strategies, and operating within the ALARA principle, the literature most often uses two terms: *reduction* and *optimization*. It is important to note that while reduction means to "reduce or diminish in size, amount, extent, or number", optimization means "an act, process, or methodology of marking something (as a design, system, or decision) as fully perfect, functional, or effective as possible" [26].

The optimization strategies that will be presented in Chaps. 3 and 5 for DR and CT respectively, will therefore focus on the specific research methodologies that address the dictionary definition of optimization.

Optimization strategies for DR and CT examinations have been identified and described in the literature and they are focused on how technical factors can be used to optimize the dose and not compromise the image quality [27]. This is the basic tenet of the ALARA philosophy.

Technical factors related to DR imaging are several and include for example, exposure technique factors (mAs and kV); the exposure indicator; the deviation index; image processing and noise reduction algorithms; and diagnostic reference levels [28–30]. Optimization strategies are therefore focused on these factors. While technical factors will be described further in Chap. 2, optimization strategies for DR imaging will be described in Chap. 3.

The image acquisition and image processing technologies are totally different in CT compared to DR imaging systems. As a result, there are significant technical factors that must be considered when describing optimization strategies in CT. Technical factors are numerous and include for example, kV; mAs; noise; pixel size; and slice thickness. Furthermore, there are several other factors that affect the dose in CT imaging. These include the pitch; the scan field-of-view; beam collimation; noise-reducing image reconstructive algorithms, particularly iterative reconstruction; for example [2, 7–9, 31–33].

While Chap. 4 will describe the essential physics and technology of CT imaging, including factors affecting the dose in CT; Chap. 5 will elaborate on specific optimization strategies that are noteworthy, and have received attention in the literature.

References

1. Hendee WR, O'Connor MK (2012). Radiation risks in medical imaging. Separating fact from fantasy. Radiology 204 (2): 312–320
2. Bushong, S. (2022). Radiologic science for technologists (12th ed.). St. Louis: Elsevier Mosby.
3. International Commission on Radiological Protection (2007). The 2007 Recommendations of the International Commission on Radiological Protection. Orlando Florida, Elsevier. ICRP Publication 103. Ann ICRP; 37: 1–332.
4. Matthews K, Brennan PC (2008). Justification of x-ray examinations: General principles and an Irish perspective. Radiography 14, 249–35

5. ICRP (2013). Radiological protection in paediatric diagnostic and interventional radiology. ICRP Publication 121. Ann. ICRP 42(2).
6. Seeram E (2019). Digital Radiography: Physical Principles and Quality Control. Second Edition, Singapore, Springer.
7. Seeram E (2023). Computed Tomography: Physical Principles, Patient Care. Clinical Applications, and Quality Control (5th Edition). St Louis, MO., Elsevier
8. Seeram E, Brennan P (2017). Radiation Protection in Diagnostic X-Ray Imaging, Burlington, MA., Jones and Bartlett Learning.
9. Bushberg JT, Seibert JA, Leidholdt EM, Boone JM (2021). The Essential Physics of Medical Imaging. 4th ed. Philadelphia, PA., Wolters Kluwer.
10. Willis CE, Slovis TL (2004). The ALARA concept in pediatric CR and DR: dose reduction in pediatric radiographic exams—a white paper conference executive summary. Pediatr Radiol; 34 (suppl 3):S162–S164
11. Seibert JA (2008), Digital radiography: image quality and radiation dose. Health Phys. Nov; 95 (5): 586–98) PubMed https://doi.org/10.1097/01.HP.0000326338.14198.a2
12. Gibson DJ, Davidson RA (2012). Exposure creep in computed radiography: a longitudinal study. Acad Radiol. Apr;19(4):458–62. doi: https://doi.org/10.1016/j.acra.12.003. Epub. Jan 5. PubMed
13. Davidson, R. (2020). Dose creep in digital radiography. In C. Hayre, & W. Cox (Eds.), General radiography: Principles and practice (pp. 59–69). CRC Press. https://doi.org/https://doi.org/10.1201/9781003047278-4
14. Benfield S, J.D. Hewis, C.M. Hayre (2021) Investigating perceptions of 'dose creep' amongst student radiographers: A grounded theory study, Radiography, Volume 27, Issue 2, Pages 605–610
15. Seibert, J. A., & Morin, R. L. (2011). The standardized exposure index for digital radiography: An opportunity for optimization of radiation dose to the pediatric population. Pediatr Radiol 41, 573–581.
16. Entesar Zawam Dalah (2020). Quantifying dose-creep for Skull and chest radiography using dose area product and entrance surface dose: Phantom study, Radiation Physics and Chemistry, Volume 167, 108231
17. Brenner DJ, Hall EJ (2007). Computed tomography: an increasing source of radiation exposure. N Engl J Med;357(22), 2277–2284.
18. Berrington de Gonzalez A, Mahesh M, Kim KP, et al (2009). Projected cancer risks from computed tomographic scans performed in the United States in 2007. Arch Intern Med;169(22):2071–2077. doi:https://doi.org/10.1001/archinternmed.440.
19. Van der Molen AJ, Stoop P, Prokop M, Geleijns J (2013). A national survey on radiation dose in CT in The Netherlands. Insights Imaging;4(3):383–390. doi:https://doi.org/10.1007/s13244-013-0253-9
20. Mathews JD, Forsythe AV, Brady Z, et al (2013). Cancer risk in 680,000 people exposed to computed tomography scans in childhood or adolescence: data linkage study of 11 million Australians. BMJ; 346:1–18. doi:https://doi.org/10.1136/bmj.f2360.
21. Hricak H, Brenner DJ, Adelstein SJ, Frush DP, et al (2011). Managing radiation use in medical imaging: a multifaceted challenge. Radiology; 258(3):889–905. doi:https://doi.org/10.1148/radiol.10101157
22. Amis ES Jr. (2011). CT radiation dose: trending in the right direction. Radiology;261(1):5–8. doi:https://doi.org/10.1148/radiol.11111319
23. FDA (2022). What are the radiation risks from CT? (https://www.fda.gov/radiation-emitting-products/medical-x-ray-imaging/what-are-radiation-risks-ct) Accessed June 2022.
24. World Nuclear Assoiation (2022). Radiation and Health Effects. https://www.world-nuclear.org/information-library/safety-and-security/radiation-and-health/radiation-and-health-effects.aspx Accessed June 2022

25. Smith-Bindman R, Wang Y, Chu P, Chung R, Einstein AJ, Balcombe J, et al (2019) International variation in radiation dose for computed tomography examinations: prospective cohort study. BMJ; 364: k4931
26. Reduction, Optimization. Merriam-Webster Dictionary. Online English Language Dictionary. www.merriam-webster.com/dictionary. Accessed June 2021.
27. Samei E, Järvinen H, Kortesniemi M, Simantirakis G, Goh C, Wallace A, et al (2018). Medical imaging dose optimisation from ground up: expert opinion of an international summit. J. Radiol. Prot. 38 967
28. Mothiram, U., Brennan, P. C., Lewis, S. J., Moran, B., & Robinson, J. (2014). Digital radiography exposure indices: A review. Journal of medical radiation sciences, 61(2), 112–118. https://doi.org/https://doi.org/10.1002/jmrs.49
29. Marcelo B. Freitas, Ricardo B. Pimentel, Laura F. Braga, Francisco S.A. Salido, Rodrigo F.C.A. Neves, Regina B. Medeiros (2020). Patient dose optimization for computed radiography using physical and observer-based measurements as image quality metrics, Radiation Physics and Chemistry, Volume 172.
30. Seeram E (2022). Dose Optimization in Digital Radiography. American Society of Radiological Technologists (ASRT), Essential Education; Pages 1–14. https://apps.asrt.org/DirectedReading/DirectedReading.aspx (accessed June 30, 2021)
31. Wolbarst AB, Capasso P, Wyant AR (2013). Medical Imaging: Essentials for Physicians. Hoboken, NJ: Wiley-Blackwell.
32. Goo HW (2012). CT radiation dose optimization and estimation: an update for radiologists. Korean J Radiol.;13(1):1–11. doi:https://doi.org/10.3348/kjr.2012.13.1.1.
33. Maldjian PD, Goldman AR (2013). Reducing radiation dose in body CT: primer on dose metrics and key CT technical parameters. AJR Am J Roentgenol.; 200(4):741–747. doi:https://doi.org/10.2214/AJR.12.9768

Chapter 2
Digital Radiography: A Technical Review

Keywords Digital radiography · Detector response to radiation · Computed radiography (CR) · Photostimulable luminescence · Flat-panel digital radiography (FPDR) · Indirect conversion digital radiography · Direct conversion digital radiography · Image processing · Standardized exposure indicator · Deviation index · Dose in digital radiography · Diagnostic reference levels

2.1 Digital Radiography: Two Notable Facts Related to Dose Management

Digital Radiography (DR) imaging systems fall into two popular imaging modalities, namely; computed radiography (CR), and flat-panel digital radiography (FPDR). These systems have been described in detail in the literature [1–3]. Furthermore, the major system components and processes common to of each these two systems were illustrated in Fig. 1.1 (Chap. 1). These include image acquisition by the detectors after exposure; image preprocessing; image postprocessing; and image display for viewing and interpretation.

The radiation transmitted through the patient creates a latent image on the detectors. The latent image is rendered visible through the use of sophisticated software. When imaging patients, operators must be cognizant of the following two notable facts which play a role in achieving the ALARA (as low as reasonably achievable) principle of the International Commission on Radiological Protection (ICRP):

1. The response of digital detectors to radiation exposure, is *linear* (Fig. 2.1) compared to a *sigmoidal* response of film-screen detectors (which are now obsolete). The consequence of a linear response is that *image processing* ensures that the density and contrast of the displayed image appear acceptable to the observer, whether the exposure is low or high. This is illustrated in Fig. 2.2. The graphs are histograms showing frequency distributions of digital code values and characteristic curves that convert digital values into optimized images that all appear with the same brightness and density. “The VOILUT is adjusted to the histogram to

E. Seeram, *Dose Optimization in Digital Radiography and Computed Tomography*,
https://doi.org/10.1007/978-3-031-22871-1_2

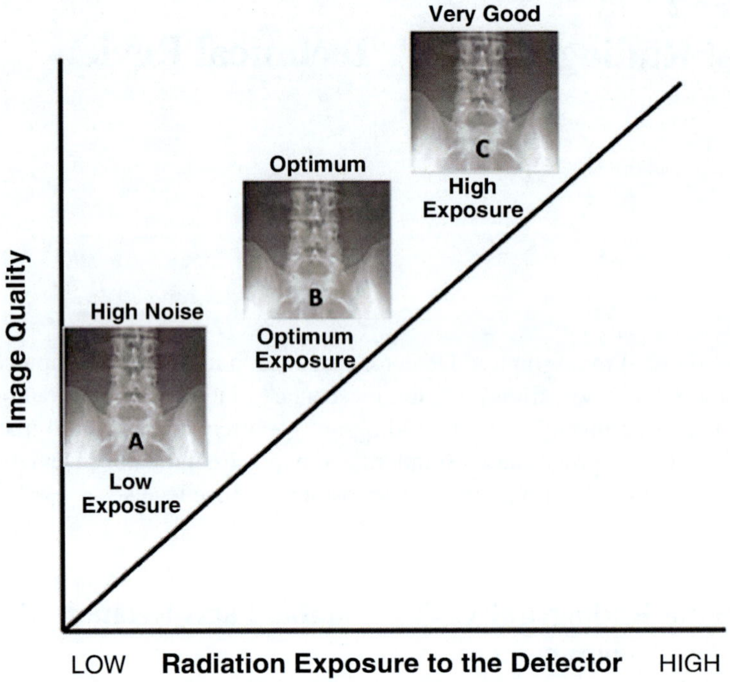

Fig. 2.1 The response of digital detectors to radiation exposure, is *linear*, compared to a sigmoidal response of film-screen detectors (which are now obsolete). (See text for further explanation)

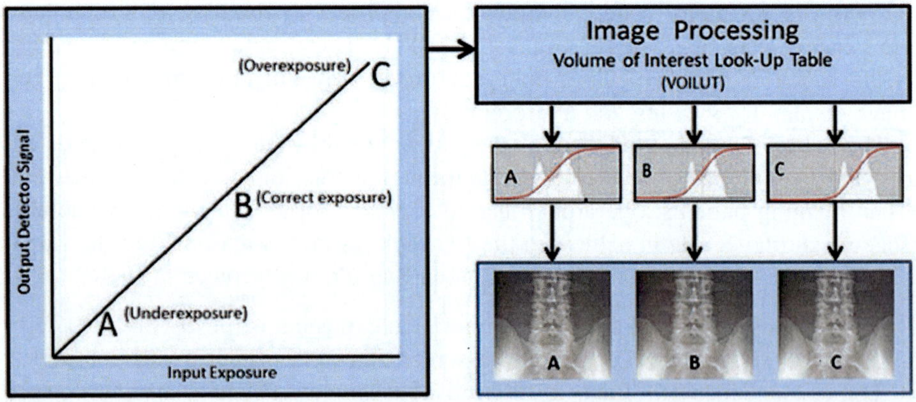

Fig. 2.2 The image processing features of DR imaging systems adjust images that are overexposed, underexposed, and correctly exposed to all appear visually acceptable in terms of image gray scale

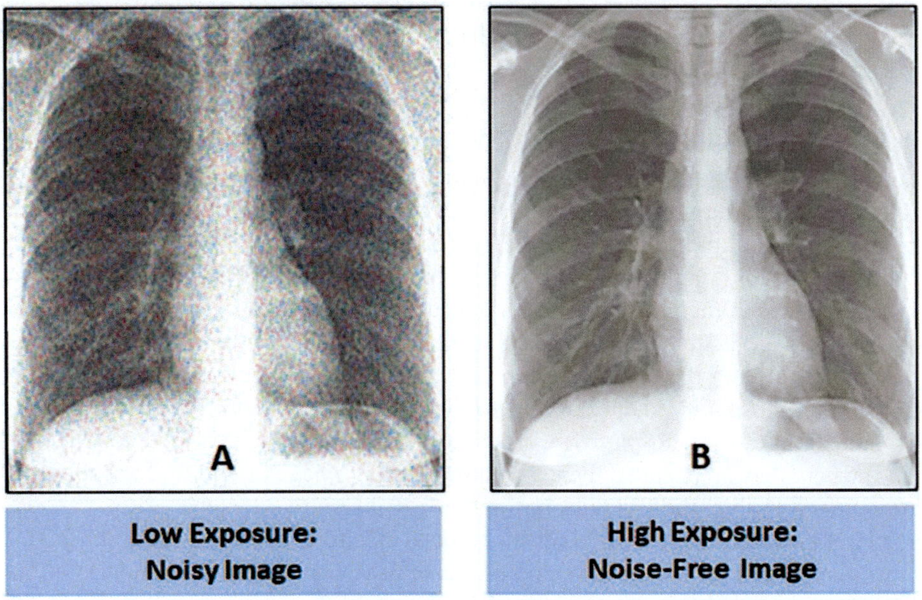

Fig. 2.3 Low exposures in digital radiography can produce noisy images (**a**); while high exposures produce excellent noise-free image (**b**); however, high exposures also increase the dose to the patient

achieve optimal rendering of the image content" [4]. Low exposure can produce noisy images (Fig. 2.3a) while high exposures produce excellent noise-free image (Fig. 2.3b); however, high exposures also increase patient exposure to radiation [1–3].

2. An *exposure indicator* (EI), a numerical value is displayed on the image to indicate that the correct exposure technique factors were used for the examination. The EI is not the dose to the patient. The more current numerical value known as the *deviation index* (DI) is now used as a cue for the exposure to the detector [4–6]

2.2 Digital Radiography Systems: A Technical Review

The physical principles and technology of DR imaging systems have been described in the literature [1–4, 7]. As noted Sect. 2.1, two popular and commonplace systems include CR and FPDR. The major system components and processes involved in creating images were illustrated in Fig. 1.1 (Chap. 1). This section will review the overall structure and function of each of these two imaging modalities.

2.2.1 *Computed Radiography: Physical Principles and Technology*

CR imaging is based on the physical principle of *photostimulable luminescence*, a characteristic of photostimulable phosphors. The phosphors typically used in CR technology are barium fluorohalides (BaFX) where X represents the halide (iodide, bromide). Another phosphor used is cesium bromide (CsBr).

Figure 2.4 shows an overview of the CR imaging process, from data acquisition (x-ray exposure of the image receptor); image processing; and image display. In the first step, image acquisition CR uses an *imaging plate* (IP) to capture radiation transmitted through the patient during the examination. The IP is coated with the appropriate phosphor. When exposed to x-rays, electrons from the valence band (ground state) of the phosphor, are raised to a higher energy level (the conduction band) where they are trapped, hence creating a latent image on the IP. To render the latent image visible, the IP is inserted into a CR image reader, where it is scanned by a laser beam. This process forces the trapped electrons in the conduction band to return to the valence band, thus creating a bluish-purple light. Such emission is called photostimulable luminescence. The light emitted is proportional to energy of the x-ray photons striking the IP. The light is captured by a photomultiplier and converts it into electrical signals. These signals are digitized by an analog-to-digital converter, and subsequently sent to the computer. In the final stage of this process, the IP is exposed to a high intensity light to remove any residual latent image (plate erasure). The IP can now be used again for another examination.

The second notable step in the CR imaging process is image processing which consists of preprocessing and post processing as shown in Fig. 2.4. Processing will be reviewed briefly in Sect. 2.3.1.

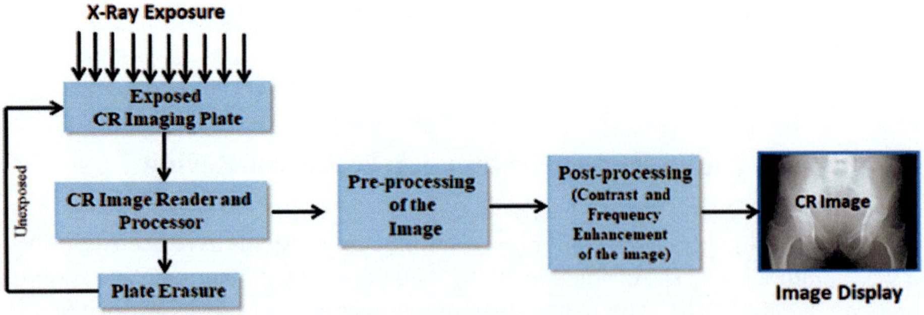

Fig. 2.4 A general overview of the CR imaging process, from data acquisition (x-ray exposure of the image receptor); to image processing; and image display. (See text for further explanation)

2.2.2 Flat-Panel Digital Radiography: Physical Principles and Technology

The shortcomings of CR, such as poor x-ray detection efficiency (affects image quality and dose), poorer spatial resolution compared to film-screen detectors, and the potential for damage to the IP as it is transported to the CR image reader for processing; have resulted in the development of other systems to solve these problems. One such system is the FPDR imaging system. The major components of a flat-panel detector and associated image processing is illustrated in Fig. 2.5. There are two types of FPDR detectors, based on how they convert x-rays to electrical signals; indirect and direct conversion detectors. For more details of the structure and function of these detectors, the reader should refer to the literature [1–3, 7]. A brief description of each of these two types of FPDR systems will be presented in Sects. 2.2.2.1 and 2.2.2.2.

Indirect Conversion Digital Radiography

Figure 2.6 illustrates the major components of a flat-panel digital radiography system based on indirect conversion detector. First, x-rays fall upon a scintillator (cesium iodide) and are converted into light using. Secondly. the emitted light from the phosphor falls on a matrix array of electronic elements consisting of an amorphous silicon (a-Si) photodiode flat-panel layer, with a thin-film transistor (TFT) to create and store electrical charges in direct proportion to the x-ray exposure. The charges produce electrical signals which are them subject to image processing to produce a FPDR image displayed on a monitor for viewing and interpretation.

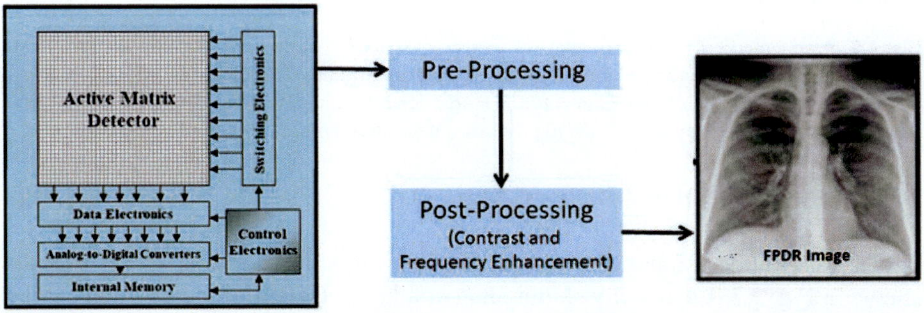

Fig. 2.5 The major components of a flat-panel detector and associated image processing. (See text for further explanation)

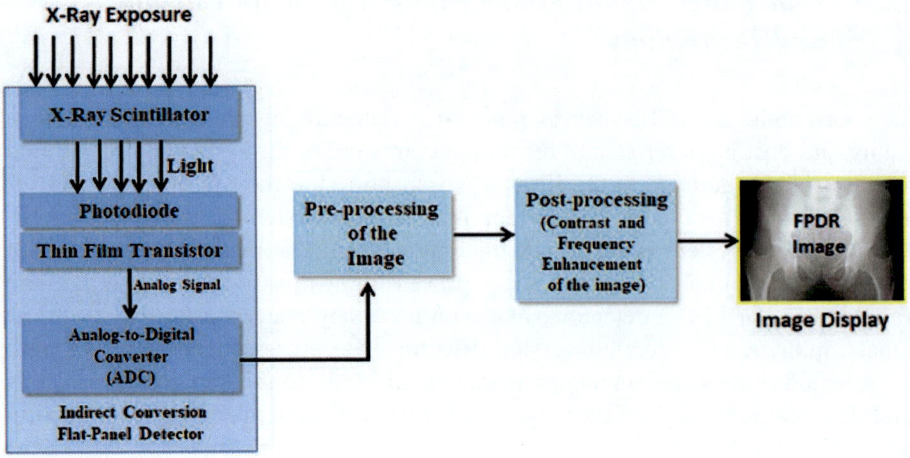

Fig. 2.6 The major components of a flat-panel digital radiography system based on *indirect conversion* detector

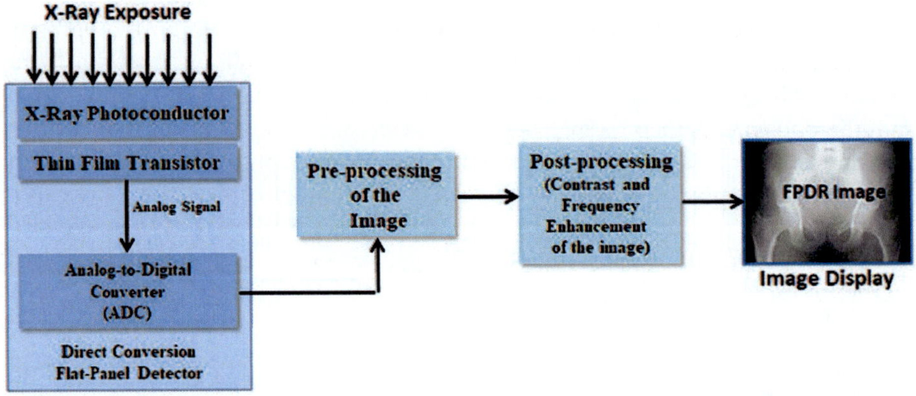

Fig. 2.7 A schematic of the components of a *direct conversion* digital radiography system. First, x-rays fall a photoconductor (selenium) and are converted directly into electrical charges that fall upon a matrix array of electronic elements and stored for subsequent readout. Secondly, the electrical signals are digitized by the ADC and processed by a digital computer to produce an image

Direct Conversion Digital Radiography

A schematic of the components of a direct conversion digital radiography system is shown in Fig. 2.7. First, x-rays fall a photoconductor (selenium) and are converted directly into electrical charges that fall upon a matrix array of electronic elements and stored for subsequent readout. Secondly, the electrical signals are digitized by the ADC and processed by a digital computer to produce an image.

2.3 Other Considerations Related to CR and FPDR Systems

Dose optimization in DR requires other considerations such as image processing, the standardized exposure indicator, and the factors affecting dose in DR. Each of these will be reviewed briefly in the appropriate sections below. Optimization strategies will be described separately in Chap. 3.

2.3.1 Image Processing

Digital image processing is an essential feature of digital radiography imaging systems, and consists of two processes, namely; image preprocessing and postprocessing which are applied to the original images. The purpose of image processing is to improve the visibility of different anatomical structures and image densities in an effort to enhance the diagnostic interpretation skills of the observer (radiologist, for example). Preprocessing and postprocessing operations have been described in previously published materials, and readers should consult the literature for more details [1–3, 7].

In digital radiographic imaging, preprocessing is mainly used to not only to identify, correct, and scale the raw image data prior to image display for the purpose of viewing and interpretation. Postprocessing on the other hand is intended to manipulate the image displayed on the viewing monitor. Such processing can be applied to sharpen and reduce noise in displayed images, as well as to alter the brightness and contrast of the image to suit the viewing needs of the observer.

One such common image processing operation is referred to as *windowing*, using controls such as the window width (WW) and the window level (WL) to change the image contrast and brightness respectively [8]. Another commonly used image postprocessing operation is *spatial frequency processing*, specifically to change the sharpness (detail) of an image.

Spatial frequency processing operates on the frequency domain illustrated in Fig. 2.8. The spatial domain image of the hand is converted into a frequency domain image using the mathematical rigor of the Fourier Transform (FT). As illustrated the frequency domain image consists of high frequencies (sharpness) and low frequencies (contrast). This processing operation is intended to change the sharpness of an image by adjusting the frequency components of the image. Using a high pass digital filter, low frequencies are suppressed to produce a sharper image, as shown in Fig. 2.9. Note the sharpness of the wrist bones compared to the original image. The use of image processing as a dose optimization strategy will be discussed in Chap. 3.

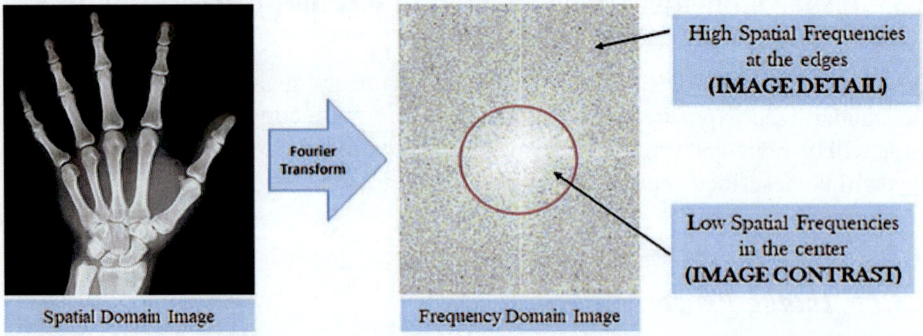

Fig. 2.8 Spatial frequency processing operates on the frequency domain as illustrated. The spatial domain image of the hand is converted into a frequency domain image using the mathematical rigor of the Fourier Transform (FT). (See text for further explanation)

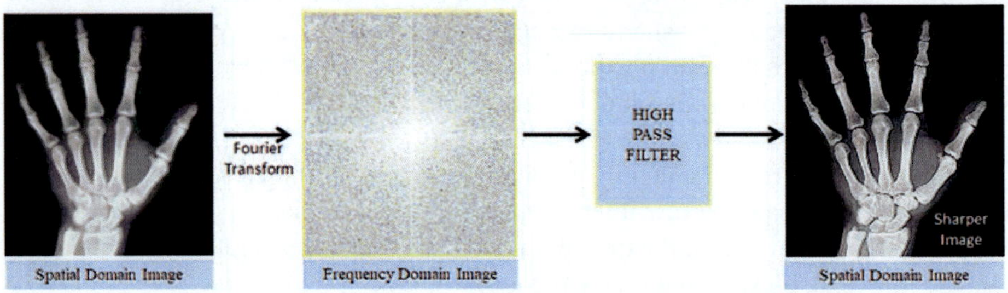

Fig. 2.9 An illustration of the use of a high pass digital filter which supresses low frequencies to produce a sharper image. (See text for further explanation)

2.3.2 The Standardized Exposure Indicator and the Deviation Index

The Exposure Indicator (EI) and the Deviation Index (DI) are numerical values that appears on the digital radiography image to indicate to the operator that the correct exposure technique was used for the examination. In the past, various vendor-specific EIs were used. Furthermore, two scales were prevalent, namely; an *inverse scale* and a *proportional scale*, as well as different algorithms and detector calibration techniques. These two scales relate EI to dose to the detector. The inverse scale states that the EI in inversely proportional to the dose (EI α 1/Dose) that is, if the detector dose is 5 μGy and the EI value for this dose is 400, then for 10 μGy and 20 μGy, the EI decreases from 200 to 100 respectively. The proportional scale states that the EI is directly proportional to the dose (EI α Dose), that is, if the EI value for detector dose of 5 μGy is 500, then the EIs for doses 10 μGy and 20 μGy, are 1000 and 2000 respectively.

These differences created some degree of confusion for operators. Subsequently, the International Electrotechnical Commission (IEC) [5] and American Association of Physicists in Medicine (AAPM) [9] in collaboration with the Medical Imaging and Technology Alliance, standardized the EI, to eliminate confusion in clinical imaging. This standardization requires that operators understand four parameters for effective use of the standardized EI in clinical practice well. These include the EI, *target EI* (EI_T), *Deviation Index* (DI), and *Volume of Interest* (VOI).

Furthermore, the IEC requires that the standardized EI uses a *linear proportional scale* related to the detector exposure/signal, that is, doubling the detector dose, doubles the standardized EI value. Additionally, there are other conditions that must be considered [5].

The IEC [5] definitions are as follows:

1. EI is a "measure of the detector response to radiation in the relevant image region of an image acquired with a digital x-ray imaging system".
2. EI_T is the "expected value of the exposure index when exposing the x-ray image receptor properly".
3. DI is a "number quantifying the deviation of the actual exposure index from a target exposure index".
4. VOI is the "central tendency of the original data in the relevant image region. The central tendency is a statistical term depicting generally the center of a distribution. It may refer to a variety of measures such as the mean, median, or the mode".

Presently all digital radiography imaging equipment use the standardized EI. The following demonstrates the linear proportional scale of the standardized EI:

- If a detector dose of 2.5 μGy generates an EI of 250, then detector doses of 5 μGy; 10 μGy and 20 μGy, will generate EIs of 500; 1000 and 2000 respectively.

Once the EI is generated, the DI which quantifies the deviation of the actual EI from the expected EI_T and is calculated using the following algebraic expression:

$$DI = 10\log_{10}\left(EI / EI_T\right)$$

The results of this calculation indicate the following [5, 9]:

- A DI number of +1 is equal to an overexposure of 26% more than the desired exposure.
- A DI number of −1 is equal to an underexposure of 20% less than the desired exposure.
- A DI number of +3 is equal to 100% more than the desired exposure.
- A DI number of −3 is equal to 50% less than the desired exposure.
- The acceptable range of DI numbers is approximately −1 to +1, and the DR system is able to deliver the EI_T established by the department.
- Furthermore, numbers greater than +1 and less than −1 would indicate gross overexposure and underexposure, respectively,

2.4 Factors Affecting the Dose in Digital Radiography: An Essential Review

There are a host of factors in digital radiography that affects the dose to the patient [1–3, 6, 10]. These include exposure factors (mAs and kV); collimation; antiscatter grids; automatic exposure control systems; source-to-skin distance; and so forth. Since the research on dose optimization in digital radiography focuses primarily on exposure factors, this section will only highlight the essential mathematical relationship between dose and mAs; and dose and kV. There are at least two important relationships for the operator to understand between dose and exposure factors, and these include:

1. The dose is directly proportional to the mAs; doubling the mAs doubles the dose [1–3, 6, 10]
2. The dose is also directly proportional to the square of the ratio of the kV. This relationship means that doubling the kV would result in an increase in x-ray intensity (dose) by a factor of 4 [2]. Furthermore, at 100 kV, the dose is 1.5 times greater than 80 kV, and 2.5 times higher at 120 kV [10]. This relationship between dose and kV demonstrates that a relatively small reduction of kV can result in a substantial reduction in total radiation dose.

The literature also shows that there are other considerations that have an impact on dose optimization. These include the EI, DI, image processing, and diagnostic reference levels. As noted by Seeram [11], "each approach has the potential to improve patient dose levels, and many of these strategies can be combined to further improve the balance of radiation dose and image quality in digital imaging". Diagnostic reference levels will be described briefly in Sect. 2.5.

2.5 Diagnostic Reference Levels

The *Diagnostic Reference Level* (DRL) is a useful tool for optimization of radiation protection used to evaluate patient doses in medical imaging [12–14]. In radiation protection, dose limits have been established by various radiation protection authorities. The ICRP for example have recommended dose limits for occupationally exposed individuals (technologists and radiologists for example), pregnant workers, and members of the public [2, 3, 6]. No dose limits have been established for patients. Therefore, the ICRP, as early as 1990, introduced the notion of DRLs to be used for patients. A significant point to note is that DRLs and dose limits are not the same thing.

2.5.1 Definitions and Major Guidelines

The ICRP and the American College of Radiology (ACR) have defined the DRL. The ICRP's definition is that DRLs "are a form of investigation level applied to an easily measured quantity, usually the absorbed dose in air or tissue equivalent material at the surface of a simple standard phantom or a representative patient" [12]. Additionally, the ICRP states that DRLs "are not for regulatory or commercial purposes, not a dose restraint and not linked to limits or constraints" [12]. The ACR has defined the DRL as "an investigation level to identify unusually high radiation dose or exposure levels for common diagnostic medical x-ray procedures" [15].

DRLs will not be described further in this text and the reader should refer to the literature [6, 12, 15] for more details. A few major guidelines from the ICRP however, are noteworthy for the purposes of this chapter. DRLs are an advisory, and not a regulatory measure, and are not related to dose limit. Their purpose is to identify high radiation doses delivered to patients using easy to measure dose quantities such as for example, skin entrance exposures. A significant feature of the DRL is that its selection is established by professional organizations using a percentile point. A notable method is to use a diagnostic reference range, where the upper limit of the range is established at the 75th percentile, and the lower limit is set at the 25th percentile of the calculated patient dose. Furthermore, while the levels below the 25th percentile may compromise image quality, levels above the 75th percentile may reflect excessive dose, and therefore the potential for dose optimization must be considered in this respect.

The ACR-AAPM-SPR [15] for example, recommends a DRLs for an adult PA chest (23 cm thickness) with grid; adult AP abdomen (22 cm thickness) and an adult AP lumbosacral spine (22 cm thickness) of 0.15 mGy; 3.4 mGy and 4.2 mGy respectively. Readers should consult the guidelines of their respective countries for established DRL values for medical x-ray examinations.

References

1. Seeram E (2019), Digital Radiography: Physical Principles and Quality Control. Second Edition, Singapore, Springer.
2. Bushong S (2021). Radiologic Science for Technologists: Physics, Biology and Protection. 12th ed. St Louis, MO: Mosby-Elsevier Inc.
3. Bushberg JT, Seibert JA, Leidholdt EM, Boone JM (2021). The Essential Physics of Medical Imaging. 4th ed. Philadelphia, PA: Lippincott Williams & Wilkins.
4. Seibert, J.A., Morin, R.L (2011). The standardized exposure index for digital radiography: an opportunity for optimization of radiation dose to the pediatric population. Pediatr Radiol; 41, 573–581. https://doi.org/10.1007/s00247-010-1954-6 PubMed

5. International Electrotechnical Commission (2008). IEC 62494-1 ed.1 Medical electrical equipment exposure index of digital x-ray imaging Systems part 1: definitions and requirements for general radiography. Geneva, Switzerland (2008)
6. Seeram E, Brennan P (2017). Radiation Protection in Diagnostic X-Ray Imaging, Burlington, MA., Jones and Bartlett Learning.
7. Carter CE, Vealé BL (2019) Digital Radiography and PACS. 3rd Edition. St Louis, MO. Elsevier, Inc
8. Seeram E (2023). Computed Tomography: Patient Care, Clinical Applications, and Quality Control. St Louis, MO. Elsevier
9. Shepard, J. S., Wang, J., & Flynn, M., et al. (2009). An exposure indicator for digital radiography: AAPM Task Group 116 (executive summary). Med Phys 36(7), 2898–2914.
10. Sharma R, Sharma SD, Pawar S, Chaubey A, Kantharia S, Babu DA (2015. Radiation dose to patients from X-ray radiographic examinations using computed radiography imaging system. J Med Phys. Jan-Mar; 40(1): 29–37. doi: https://doi.org/10.4103/0971-6203.152244
11. Seeram, E (2022). Dose Optimization in Digital Radiography. Radiologic Technology; Directed Readings Website, American Society of Radiologic Technologists (ASRT), New Mexico
12. ICRP Publication 135 (2017). Diagnostic Reference Levels in Medical Imaging. Annals of the ICRP Volume 46, Issue 1, October, Pages 1–144
13. Seeram, E. and Brennan, P (2006). Diagnostic Reference Levels in Radiology. Radiologic Technology, May/June 2006, Vol. 77/No. 5; 373–384
14. Seeram, E. (2020). Rad Tech's Guide to Radiation Protection, 2e. Oxford: Wiley, 2020
15. ACR-AAPM-SPR (2018) Practice Parameter for Diagnostic Reference Levels and Achievable Doses in Medical X-Ray Imaging. https://www.acr.org/-/media/ACR/Files/Practice-Parameters/Diag-Ref-Levels.pdf Accessed June 2022.

Chapter 3
Optimization Strategies in Digital Radiography

Keywords Dose optimization in digital radiography · Exposure technique factors · kV · mAs · Exposure indicator · Deviation index · Image post processing · Multifrequency processing · Noise reduction algorithms · Diagnostic reference levels

3.1 Introduction

As reviewed in Chap. 1, optimization is a radiation protection principle, first provided by the International Commission on Radiological Protection (ICRP) [1] and now is a mandatory requirement of various radiation protection organizations around the world. In summary, during the conduct of the examination, optimization entails not only adjusting the "imaging parameters" [1], but also to "institute protective measures" [1]. These two tasks are intended to ensure that the first part of optimization, the ALARA (as low as reasonably achievable) principle is observed. The second part of optimization, as noted by the ICRP [2], involves at least four elements, namely; the imaging equipment, the adequacy of the imaging equipment, technical parameters of the examination, and diagnostic reference levels.

This Chapter will outline several strategies that have been used successfully in digital radiography, to optimize the dose to the patient without compromising image quality needed for diagnosis. These strategies include optimization of the *exposure technique factors* (mAs and kV); the *Exposure Indicator* (EI); the *Deviation Index* (DI); *image post processing algorithms*; and finally, the use of *diagnostic reference levels* [3]. As noted by Seeram [3], "each approach has the potential to improve patient dose levels, and many of these strategies can be combined to further improve the balance of radiation dose and image quality in digital imaging".

E. Seeram, *Dose Optimization in Digital Radiography and Computed Tomography*,
https://doi.org/10.1007/978-3-031-22871-1_3

3.2 Optimization of Exposure Technique Factors

The evolution and continued development of digital radiography imaging systems can be traced back to the introduction of Computed Radiography (CR) imaging systems to replace film screen radiography (FSR). The limitations imposed by CR lead to the introduction of Flat-Panel Digital Radiography imaging systems which includes indirect conversion imaging systems and direct conversion imaging systems [4]. FSR has now been replaced by digital radiography imaging systems.

The correct use of these digital imaging systems in clinical practice provided the motivation conduct research on several areas of performance. A significant topic for research in the early stages of use focussed on the correct exposure technique factors to use for various examinations, since users were well versed in exposure technique factors for FSR. Section 3.3 of this Chapter will explore the optimization of exposure technique factors.

3.2.1 Optimization of kV and mAs

One of the earliest reviews entitled "*Optimization research exposure technique approaches in CR imaging*" [5] assessed three major optimization approaches, namely, optimization of kV; optimization of mAs; and optimization of the Exposure Indicator (EI) in practice. This section will focus on the kV and mAs optimization approaches. The EI optimization will be discussed in Sect. 3.2.

The basis for exploring the *optimization of the kV in CR imaging* relates to the response of the CR detector to the beam energies used in diagnostic radiology compared to that of FSR detector [6].

The k-edge for a typical CR phosphor such as BaFBr/I (barium fluorobromide/iodide) is about 37 keV (compared to FSR detector such as Gd_2O_2S which has a k-edge of about 50 keV). Moore et al. suggest therefore "that lower tube voltages relative to those used for film-screen should be used for CR" [6]. This physics fact provided the motivation for research on optimization of the kV in CR imaging. With this in mind, the literature is replete with studies focusing on optimization of kV using different CR imaging systems and different body parts to the extent that these studies have produced "conflicting results" [6, 7]. This was followed by additional research to include indirect and direct flat panel detectors.

A study by Sun et al. [8] on "*Optimization of chest radiographic imaging parameters: a comparison of image quality and entrance skin dose for digital chest radiography systems*", the researchers explored the use of kV values of 100, 110, and 120, and mA settings of 1, 2, 4, 8, and 10. The found that a reduction of kV from 120 to 100 kV resulted in a dose reduction of up to 44% and without compromising image quality. A recent study by Moore et al. [9] investigated the use of an optimum range of kV for imaging the abdomen and pelvis. Their results showed that while for the abdomen a range of 70 to 120 kV was optimum, a range of 80 to 120 kV was the optimum for imaging the pelvis. Furthermore, these investigators also demonstrated that an optimum range for imaging the lumbar spine was of 80 to 120 kV [9]. These studies suggest that the kV can be reduced without affecting the image quality necessary for diagnosis [8, 9].

The need to explore *optimization of the mAs* in digital radiography stems from the fact that "manipulation of the operating kV cannot stand alone even with digital systems, and concomitant compensation of the applied mAs, together with adequate scatter control are necessary" [10]. This notion suggests that optimization must consider evaluating image quality with respect to the dose per image. An increase in the dose per image will reduce the noise, hence image quality improves [11]. Optimization should therefore consider the dose per image in assessing image quality.

The literature suggests that this approach, that is using the lowest mAs settings "without excessive noise degradation of image quality" in optimizing the dose-image quality in CR Imaging [10]. In this regard, Fauber et al. [7] investigated the "insufficient and excessive radiation exposure on CR image quality", using five images at exposure groups of 1 mAs; 2 mAs; 4 mAs; 8 mAs (baseline value); 16 mAs; 32 mAs; 64 mAs; and 125 mAs. The researchers explained that "the mAs value to achieve an optimal quality QC phantom image (line pairs) was ultimately decreased by 300% for the lowest mAs value and increased 400% for the highest mAs value". An important finding of this study is that at low exposures it was difficult to see the line pairs/mm compared to the baseline value of 8 mAs, and image mottle (noise) was pronounced at low exposures below the baseline value of 8 mAs.

The above studies are intended to set the stage for further research into the optimization of exposure technique factors, as digital radiography detectors continue to be developed with the goal of improving performance in clinical imaging of patients.

3.3 Optimization of the Exposure Indicator

As noted by Seeram et al. [5], "Optimization of the EI is closely linked with optimization of exposure technique factors (kVp and mAs)". The EI was described in detail in Chap. 2 [3, 4, 5, 12].

In summary, the EI is a numerical parameter devised by CR manufacturers to provide the user with a visual cue as to the amount of exposure to the CR IP. It provides the technologist with an indication of whether the appropriate exposure technique factors are used for the particular x-ray examination. It is quite important to understand that the *EI is not the patient dose*. Patient dose depends on several factors such as kV, mAs, beam filtration, Source-to-Image Receptor Distance (SID), body part being imaged, and collimation, for example.

The literature on optimization of the EI is extremely sparse in this area of research. Two notable early studies that set the pace for further research in this context are one by Peters and Brennan [13], and the other by Warren-Forward [14].

While the former study addressed the notion of an optimized EI as a dose management strategy via establishing an optimized mAs; the latter sought to explore the use of the EI by radiographers in clinical practice, as well as exposure creep, and the relationship between EIs and radiation dose. The fundamental conclusions of these two studies are as follows:

1. Peters and Brennan [13], reported that "although dose levels to the patient have not been directly measured, it is safe to assume that a strong correlation exists between higher exposure indices and higher patient dose".
2. Warren-Forward [14], (who used two Kodak CR systems) reported that "a small increase in the EI produced a large increase in the entrance surface dose .....an EI of 2000 produced at 125 kVp can deliver the same patient dose as an EI of 1700 produced at 70 kVp, where the EI difference of 300 represents a doubling of the dose to the detector"

The results of these two studies prompted ideas for future investigations and in this regard, Fauber et al. [7], to conclude in their study that "further research is necessary to understand the significance of the exposure indicator and its relationship to exposure techniques and patient exposure".

To explore the optimization of the EI, Seeram et al. [15], designed a noninterventional descriptive correlational research study to correlate the EI with the *entrance skin dose* (ESD) using a range of mAs values, while holding the kilovoltage peak (kV) constant. The study compared the optimized EI with the manufacturer's recommended values for the antero-posterior (AP) pelvis and AP lumbar spine projections.

The methodology used in this study, involved the following steps using the ESD and EI as dependent variables, and mAs as the independent variable:

1. The ESD for the AP pelvis and the AP lumbar spine using the digital radiography vendor's (Fuji) suggested exposure technique factors including mAs and kV.
2. The ESD measurements were measured with a calibrated dosimeter (Unfors Instruments) free-in-air (ie, an air environment free of any absorbing or scattering objects) for the AP pelvis and the AP lumbar spine.
3. The mean milligray (mGy) per mAs value was calculated for all mAs values for the AP pelvis and AP lumbar spine.
4. Descriptive statistics (eg, such as sample size, mean, standard deviation, and range) were used for the dosimetry and EI data.
5. The Pearson correlation was applied to examine the relationship between the ESD and the EI as well as the ESD and mAs.
6. The results were graphed and plotted showing the mean dose for the AP pelvis and the AP lumbar spine as a function of mAs (Fig. 3.1); the inverse EI; the mAs; and the doses were compared. The inverse EI shows a strong positive linear relationship ($r = 0.999$) for both the AP pelvis and the AP lumbar spine.
7. At doses of 5 mGy, 10 mGy, and 20 mGy, the EIs were 400, 200, and 100, respectively.

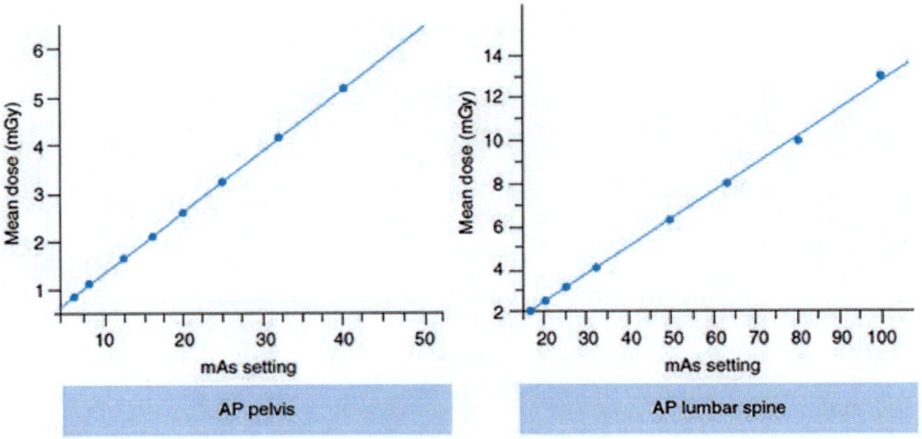

Fig. 3.1 Graphs of the mean dose in mGy plotted as a function of mAs, for the AP pelvis and the AP lumbar spine

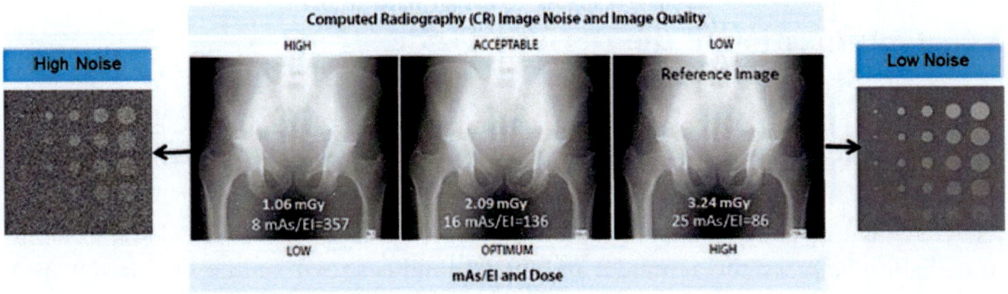

Fig. 3.2 The Reference image of the AP pelvis obtained at 25 mAs with a dose of 3.4 mGy. The optimized image recorded at a reduction of 16 mAs at 2.09 mGy, without loss of image quality. The noise test tool on the right and left of the image set shows the visual appearance of the mottle or graininess of the images

Furthermore, the dosimetry phase of this experiment showed a strong positive linear relation-ship (r = 0.999) between mAs and dose, mAs and the inverse EI, and the inverse EI and dose for the AP pelvis and AP lumbar spine. It is important to note that under the controlled conditions used in this study, the EI values were stable. The vendor's reference values of 25 mAs (EI = 86) for the AP pelvis and 50 mAs (EI = 88) for the AP lumbar spine {were optimized to 16 mAs for the AP pelvis (EI = 136)}as shown in Fig. 3.2; and 32 mAs for the AP lumbar spine (EI = 139), as shown in Fig. 3.3.

Expert observers' assessment of image quality showed that the vendor's recommended dose can be reduced by 36% for the AP pelvis and AP lumbar spine without compromising image quality.

This study clearly demonstrates that operators can optimize the dose and image quality in computed radiography imaging system using the mAs and associated EIs.

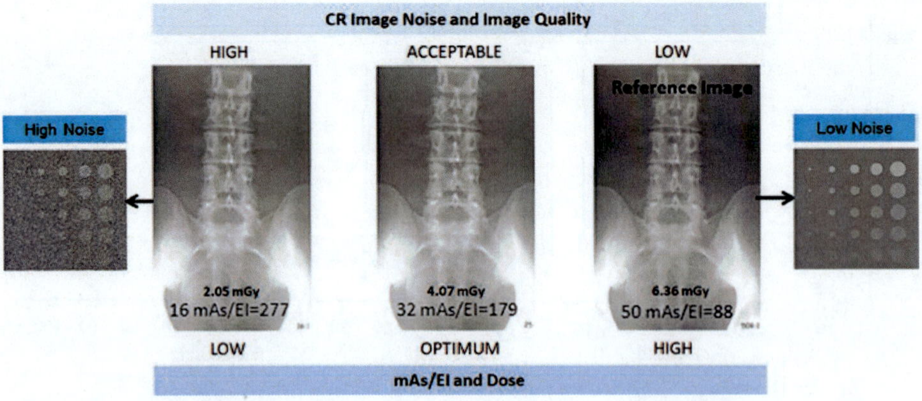

Fig. 3.3 The Reference image of the AP lumbar spine obtained at 50 mAs with a dose of 6.36 mGy. The optimized image recorded at a reduction of 32 mAs at 4.07 mGy, without loss of image quality. The noise test tool on the right and left of the image set shows the visual appearance of the mottle or graininess of the images

3.4 Optimization of the Deviation Index

The *deviation index* (DI) is an important parameter in the standardized EI paradigm, and it has been described in detail in Chap. 2. In summary, the DI is related to a target EI (EI_T) which must be established by the imaging department using the ALARA (as low as reasonably achievable) principle. The DI indicates the amount of deviation between the actual EI and the EI_T and is shown on the final image as a measure of dose optimization.

The literature is sparse on research studies on the use of the DI as an optimization strategy, however, only three studies will focus attention on the notion of using the DI as an optimization strategy in digital radiography with the ALARA principle in mind. The first study is one entitled "*Optimising default radiographic exposure factors using Deviation Index*" by Creeden and Curtis [16], and sets the stage for further research on using the DI. Having performed an analysis of the exposure logs for six DR units on five different examinations, the results showed that by reducing the mAs values for these examinations were reduced by between 22% and 50% using "multiple optimisation cycles the number of examinations outside the vendor's suggested optimal range had decreased by 28.4%". The researchers concluded that there was "a marked reduction in patient doses can be achieved through a departmental programme of DI value monitoring and targeted optimisation of default exposure settings" [16].

The second study is one examining the efficient use of the DI in optimizing dose using keywords such as "dose indicator; dose optimization; and radiation exposure" [17]. The results showed inefficient use of both the EI_T (target EI set by the imaging department) and the DI. This study provides the motivation for yet another investigation of the use of the DI in optimizing the dose in digital radiography, since the researchers point out that the implications for practice is that "Current recommendations on targeting the mean DI of 0 need to be reinforced. Theoretical knowledge and training need to be improved" [17].

The third and final study in this chapter, is one is one by Park et al. [18], that examined the use of the EI, EI_T and DI as tools for patient dose optimization/Part of the methodology involved analyzing how the DI changes with EI_T established by the department, based on the Korean national DRLs. for posterior-anterior chest, a lateral and anterior-posterior of the abdomen in clinical practice. The conclusion of this study suggests that "as the exposure conditions and DRLs varied, the clinical EI, EI_T and DI also varied. These results reveal that the clinical EI, EI_T and DI can be used as tools for optimizing the patient dose if EI_T is periodically and properly updated" [18].

In addition to the above strategies, the use of image processing techniques and noise reduction algorithms (to improve visualization of not only various anatomical structures but also pathology); have been explored as optimization strategies, adding to the traditional optimization strategies for digital radiography.

3.5 Optimization Using Image Postprocessing Algorithms

In Chap. 2, image processing was described briefly to set the stage for using it as an optimization tool in digital radiography. Image processing is a complex topic and will not be described in any detail in this book, however, an understanding of the following aspects of image processing are essential to aid its use as an optimization strategy: Why image processing of digital radiography images? And two major processes of image processing.

Image processing for Radiologic Technologists has been described in the literature [19, 20]. This topic is now a part of radiologic technology curriculum that ii has been described by Seeram in a textbook on Digital Radiography [4]. Image processing in Digital Radiography consists of two processes, namely *preprocessing* and *postprocessing*, as illustrated in Fig. 3.4.

Preprocessing operations are used to identify, correct, and scale the raw image data apply appropriate corrections, such as artifact corrections for example, to the

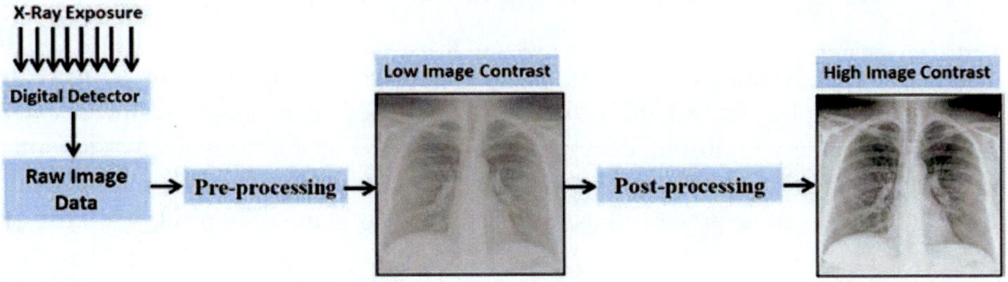

Fig. 3.4 Image processing in digital radiography consists of two processes, namely preprocessing and postprocessing. (See text for further explanation)

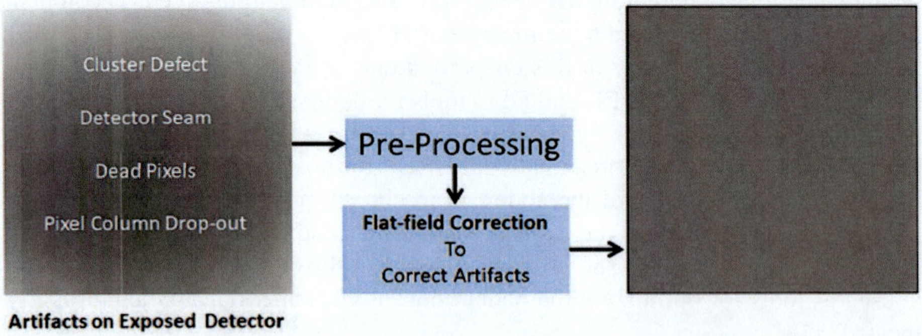

Fig. 3.5 The overall goal of an algorithm referred to as *Flat-Field Correction* is to correct for artifacts in the raw data before the image is displayed for viewing by an observer

raw data. Each DR vendor offers proprietary algorithms for their systems and therefore, it is not within the scope of this book to describe these algorithms. One such algorithm however, is a standard calibration operation known as *flat-field correction* as illustrated in Fig. 3.5.

Postprocessing, on the other hand consists of a wide range of image processing operations intended to change the image contrast, reduce image noise and enhance the sharpness of the image displayed in an effort to enhance diagnostic interpretation. These include point processing operations such as grayscale processing (windowing, image subtraction, and temporal averaging), local processing operations (such as, spatial filtering, edge enhancement, and smoothing), and global operations such as the Fourier transform (FT) [4, 21, 22]. The use of the FT coupled with a high pass digital filter to sharpen an image is shown in Fig. 2.9. Examples of noise control image postprocessing algorithms include Fujifilm Flexible Noise Control (FNC) and Multi-objective Frequency Processing (MFP); Philips Healthcare-UNIQUE (UNIfied Image QUality Enhancement); and the advanced noise reduction image processing engine (S-Vue™; Samsung Healthcare).

3.5.1 Optimization Studies Using Multifrequency Processing and Noise Reduction Algorithms

There are several examples of image postprocessing as an optimization strategy in digital radiography [3] and the general findings show that image postprocessing algorithms, namely multifrequency processing can be successfully used as an optimization strategy by demonstrating that a low-dose image with poor image quality (eg, high noise) can be processed using specialized software to improve visibility of structures [23–27].

This sub-section of this chapter will review briefly the studies by Precht et al. [23]; Lee et al. [25]; Feghali et al. [26]; and by Oh et al. [27], as summarized in an article in *Radiologic Technology: Essential Education Directed Readings* article by Seeram [3].

The American Association of Physicists in Medicine (AAPM) [24], explains that multifrequency processing software is designed to divide an image into separate frequency ranges for processing before they are recombined into an enhanced output image. In this subsection, a number of direct quotes will be cited from the research literature, so as not to detract from the investigators' original meaning.

As early as 2012, Precht et al. [23], explored the use of multifrequency processing as an optimization strategy in digital radiography in pediatric radiographic imaging. Using an anthropomorphic phantom, a total of 110 digital radiography images were obtained with a Canon DR imaging system with a CXDI-50 C detector and MLT[S] software. Furthermore, 3500 images taken of a contrast-detail phantom (CDRAD 2.0- Artinis Medical Systems [23]) were obtained. While the former set of images were assessed by three pediatric radiologists using Visual Grading Analysis, an image quality assessment tool, the latter image sets were evaluated to provide an objective image-quality assessment. The results of this investigation showed that "optimal image-quality was maintained at a dose reduction of 61% with a multi-Laplacian transformation sub-band pyramidal enhancement-technology [MLT(S)] optimized images. Even for images of diagnostic quality, MLT(S) provided a dose reduction of 88% as compared to the reference image. Software impact on image quality was found significant for dose (mAs), dynamic range dark region and frequency band" [23]. The researchers concluded that by optimizing image processing parameters, a significant dose reduction is possible without significant loss of image quality [23].

In another study utilizing multifrequency processing as an optimization strategy is one by Lee et al. [25], entitled "*Radiation dose reduction and improvement of image quality in digital chest radiography by new spatial noise reduction algorithm*". The researchers used an advanced noise reduction image processing engine (S-Vue™-Samsung Healthcare). Figure 3.6 illustrates the basic comparison between a conventional algorithm (S-Vue™ 3.00) and new image processing algorithm (S-Vue™ 3.02). The methodology involved 69 patients, each of whom was exposed to a baseline dose of 4.2 μGy using conventional image processing engine. Secondly, images were obtained using a dose reduction protocol of 2.5 μGy; 2.11 μGy; and 1.78 μGy using the new image processing engine. Images were assessed by two thoracic radiologists on "radiolucency of unobscured lung, pulmonary vascularity, trachea, edge of rib, heart border, intervertebral disc space, and pulmonary vessels in the retrocardiac area" and scored on a 5-point scale of these anatomical landmarks. The results showed that "the image quality of the low-dose image was not inferior to that of the baseline dose image even if the maximum average dose reduction rate was reduced to 47.8% of the baseline dose" [25]. Example images with optimized doses are shown in Figs. 3.7 and 3.8.

In 2021, Feghali et al. [26], published a dose optimization study in Diagnostic Interventional Imaging entitled "*New image quality and dose reduction technique for pediatric digital radiography*". In this study, the S-Vue™ (Samsung Healthcare) advanced noise reduction image processing engine was used. Having obtained images from 174 pediatric patients; "artificial noise was added to the images to simulate acquisitions at 50%, 32% and 12.5% of the routine dose levels.

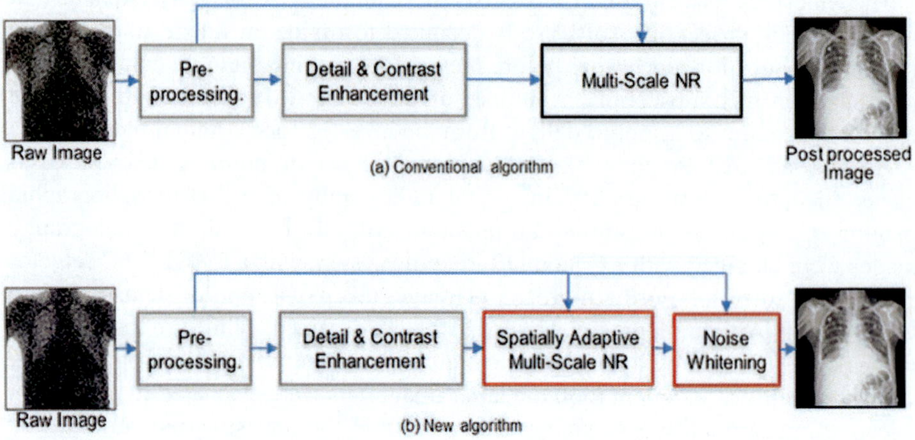

Fig. 3.6 A schematic illustration comparing the conventional and new algorithms of a dose optimization study. (Reproduced with permission under the Creative Commons Attribution 4.0 International license. Lee W, Lee S, Chong S, Lee K, Lee J, Choi JC, et al. Radiation dose reduction and improvement of image quality in digital chest radiography by new spatial noise reduction algorithm. PLOS ONE. 2020; 15 (2): e0228609. creativecommons.org/licenses/by/4.0/)

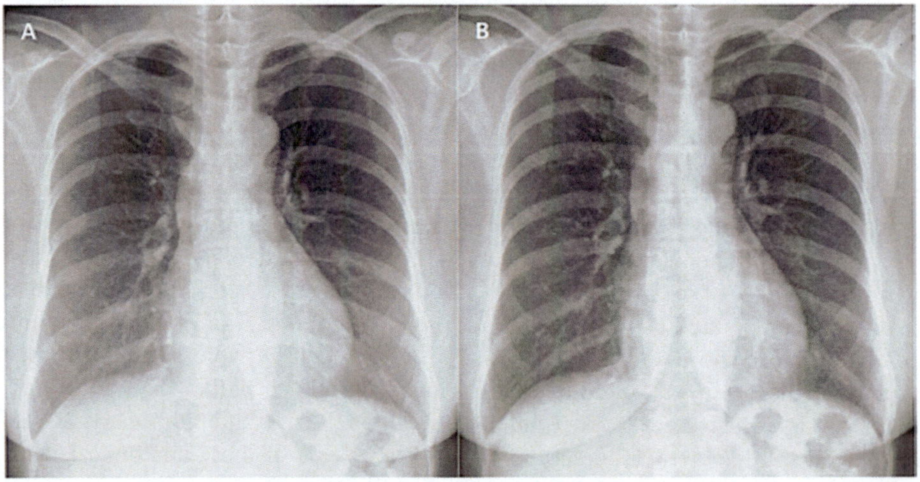

Fig. 3.7 Chest radiographs demonstrating a well-defined, small nodule in right upper lung(arrows). A. Base-line image with a standard dose of 61.9 μGy ESE. B. Optimized, low-dose image of 30.2 μGy ESE. (Reproduced with permission under the Creative Commons Attribution 4.0 International license. Lee W, Lee S, Chong S, Lee K, Lee J, Choi JC, et al. Radiation dose reduction and improvement of image quality in digital chest radiography by new spatial noise reduction algorithm. PLOS ONE. 2020; 15 (2): e0228609. creativecommons.org/licenses/by/4.0/)

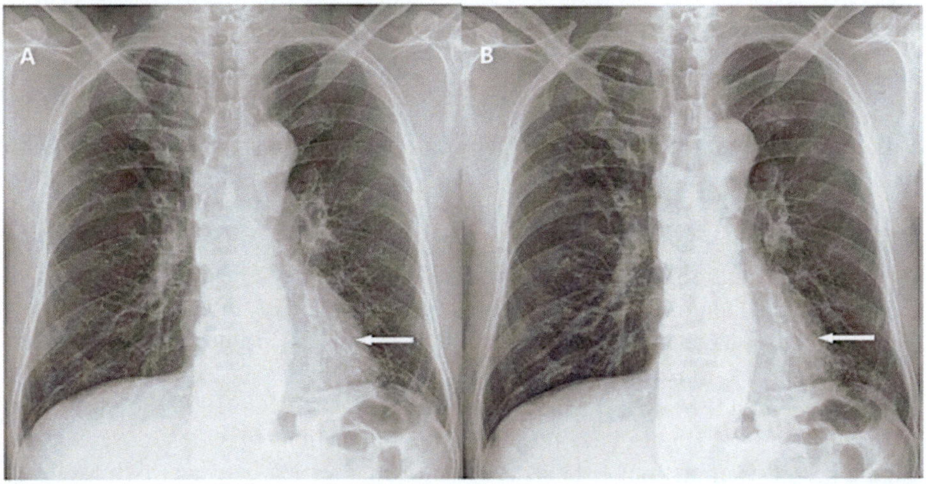

Fig. 3.8 The two chest radiographs show a subsegmental consolidation in a retrocardiac region of a left lower lung zone (arrows). For image A, the radiation dose of ESE was 54.2 μGy. The optimized image in B recorded with 25.0 μGy, which was reduced to 46.1% of the baseline dose. (Reproduced with permission under the Creative Commons Attribution 4.0 International license. Lee W, Lee S, Chong S, Lee K, Lee J, Choi JC, et al. Radiation dose reduction and improvement of image quality in digital chest radiography by new spatial noise reduction algorithm. PLoS ONE. 2020; 15 (2): e0228609. creativecommons.org/licenses/by/4.0/)

A total of 696 images corresponding to four dose levels were post-processed using S-Vue™ and further blindly scored by three pediatric radiologists using a scoring grid of 4-6 criteria specifically defined per anatomical area" using appropriate statistical analysis [26]. The results showed the use of the S-Vue™ image post-processing engine (software) reduced the dose by a factor of two, without compromising the diagnostic image quality in pediatric digital radiographic imaging.

A fourth dose optimization study in digital radiography is one by Oh et al. [27], using an advanced spatial noise reduction algorithm (ASNR) in pediatric digital radiography. Having obtained 9 sets of 30 images at "different levels of low-dose image sets of these 270 images were generated by a noise simulation tool after validation testing using phantoms. Each image set was obtained with both the ASNR and conventional algorithm, and grouped randomly and blinded". Image assessment required that each of three radiologists select the "image with optimum dose" keeping the ALARA principle in mind. The results showed that the "average of the calculated ESE (entrance skin exposure) was lower with the ASNR algorithm than with the conventional algorithm group" demonstrating that the ASNR algorithm can be a useful tool in dose optimization in digital radiography of pediatric patients.

3.6 Diagnostic Reference Levels: A Useful Tool in Dose Optimization

A recent article by Damilakis and Vassileva [28] *entitled "The growing potential of diagnostic reference levels as a dynamic tool for dose optimization*" and published in *Physica Medica*, pointed out that Diagnostic Reference Levels (DRLs) are useful tools in optimization od radiation protection in medical imaging. DRLs have been identified as useful tools in radiation protection of patients in medical imaging [29] and as such, the literature is replete with articles on the DRLs, in terms of what they are; why they are needed; how to establish DRLs; and DRL values for various digital radiography examinations.

The concept of the DRL was first introduced by the ICRP in 1990 [30], based on the fact that several studies have revealed large fluctuations in patient doses for the same types of examinations [31]. These fluctuations varied between high levels to low levels of radiation doses. DRLs are intended to focus attention on how to address these variations. It is not within the scope of this chapter to describe details of the DRL, such as for example how DRLs are established, and readers should refer to the ICRP publication of DRLs [30]. The following summary points are essential for a reasonable understanding of the DRL:

1. The *definition of the DRL* has been provided by the ICRP and the American College of Radiology (ACR). While the ICRP definition states that DRLs "are a form of investigation level, applied to an easily measured quantity, usually the absorbed dose in air, or tissue-equivalent material at the surface of a simple standard phantom or a representative patient" [30], the ACR defines a DRL as "an investigation level to identify unusually high radiation dose or exposure levels for common diagnostic medical x-ray procedures." [32]. The DRL is seen as an advisory measure and not a regulatory one, and it is not related to dose limits established for occupationally-exposed individuals (radiation workers), and members of the public.
2. When dealing with the optimization of radiation protection, the DRL addresses the issue of patient dose, and it is intended to identify high patient doses for various common examinations using specific imaging equipment. It is also recommended that when conducting a DRL study, the dose quantities and procedure used to measure the doses should be easy (simple). The entrance skin exposure (ESE) is a suggested dose quantity for example.
3. Professional organizations, such as the AAPM, ACR, the National Council on Radiation Protection and Measurements (NCRP), the Canadian Association of Radiologists (CAR) in North America for example, should establish DRLs, using a percentile point on the observed distribution for patients, and specific to a country or region. One pragmatic method is to use what has been referred to as a diagnostic reference range. While the upper level of the range is established at 75th percentile, the lower level is set at 25th percentile of the calculated patient dose. Below a 25th percentile level, image quality may be compromised, above the 75th percentile may indicate excessive dose.

4. For actual DRL values for various imaging examinations, the reader should consult their respective national radiation protection organizations. The ACR for example provides DRLs for the PA chest and AP abdomen to be 0.22 mGy and 5.3 mGy respectively [32].

As described in the above sections, optimization in digital radiography involves a number of significant strategies such as optimization of the exposure technique factors (kV/mAs); exposure indicator; deviation index; use of multifrequency processing and noise reduction algorithms; and finally, the use of DRLs as a tool in the optimization of radiation protection.

Following a literature review not only to "identify the reasons why optimization is important and the steps to be followed for a successful optimization process in digital radiology, Computed Tomography, interventional radiology and mammography", Tasapaki [29] proposed that there are five steps to a dose optimization process, including:

1. "establishment of a quality assurance programme; a mistake, misuse or malfunction of an X-ray machine can potentially affect the health or life of thousands of people,
2. establishment of a dose optimization team consisting of a radiologist, medical physicist and radiation technologist,
3. determination of baseline dose levels and image quality as well as comparisons with benchmarks to decide which exam protocols should be optimized,
4. modification of protocols by the medical physicist and,
5. evaluation of the optimization process and its effect on patient dose and image quality" [29].

It is important to stress that in the optimization of dose and image quality in medical imaging, the technologist must be an active participant in the process.

References

1. International Commission on Radiological Protection (2007). The 2007 Recommendations of the International Commission on Radiological Protection. Orlando Florida, Elsevier. ICRP Publication 103. Ann ICRP; 37: 1–332.
2. ICRP (2013). Radiological protection in paediatric diagnostic and interventional radiology. ICRP Publication 121. Ann. ICRP 42(2).
3. Seeram E (2022). Dose Optimization in Digital Radiography. American Society of Radiological Technologists (ASRT), Essential Education; Pages 1–14. https://apps.asrt.org/DirectedReading/DirectedReading.aspx (accessed June 30, 2022)
4. Seeram E (2019). Digital Radiography: Physical Principles and Quality Control. 2nd Edition. Springer Nature Singapore Pte Ltd. https://doi.org/10.1007/978-981-13-3244-9
5. Seeram E, Davidson R, Bushong S, Swan H (2013). Radiation dose optimization research: Exposure technique approaches in CR imaging: A literature review, Radiography; 19: 331e338
6. Moore CS, Saunderson JR, Beavis AW (2008). Investigating the exposure class of a computed radiography system for optimisation of physical image quality for chest radiography. The British Journal of Radiology; 82(981):771e7

7. Fauber TL, Cohen TF, Dempsey MC (2011). High kilovoltage digital exposure techniques and patient dosimetry. Radiologic Technology; 82(6):501e9.
8. Sun Z, Lin C, Tyan Y, Ng KH (2012). Optimization of chest radiographic imaging parameters: a comparison of image quality and entrance skin dose for digital chest radiography systems. Clin Imaging. 2012 Jul–Aug;36(4):279–86. doi: https://doi.org/10.1016/j.clinimag.2011.09.006 Epub; Jun 8. PubMed
9. Moore CS, Wood T, Balcam S, Needler L, Guest T, Ngu WP, Chong LW, Saunderson J, Beavis A (2020). Optimisation of tube voltage range (kVp) for AP abdomen, pelvis and spine imaging of average patients with a digital radiography (DR) imaging system using a computer simulator. Br J Radiol. Oct 1;93(1114):20200565. Doi: https://doi.org/10.1259/bjr.20200565 Epub 2020 Aug 12. PubMed; PMCID: PMC7548356
10. Matthews K, Brennan PC (2009). Optimisation of X-Ray examinations: General principles and an Irish perspective. Radiography; 15:262e8.
11. Marshall NW (2001). Optimisation of dose per image in digital imaging. Radiation Protection Dosimetry; 94 (1e2):83e7.
12. Seibert JA (2008). Digital radiography: image quality and radiation dose. Health Phys. Nov;95(5):586–98) PubMed https://doi.org/10.1097/01.HP.0000326338.14198.a2
13. Peters SE, Brennan PC (2002). "Digital radiography: are the manufacturers" settings too high? Optimisation of the Kodak digital radiography system with the aid of the computed radiography dose index. European Radiology; 12 (9):2381e7.
14. Warren-Forward H, Arthur L, Hobson L, Skinner R, Watts A, Clapham K, et al (2007). An assessment of exposure indices in computed radiography for the posterioranterior chest and the lateral lumbar spine. The British Journal of Radiology; 80 (849):26e31.
15. Seeram E, Davidson R, Bushong S, Swan H (2016). Optimizing the Exposure Indicator as a Dose Management Strategy in Computed Radiography. Radiologic Technology, March/April, Volume 87, Number 4; 380–391 PubMed
16. Creeden A, Curtis M (2020). Optimizing default radiographic exposure factors using Deviation Index. Radiography; 26: 308–313 PubMed https://doi.org/10.1016/j.radi.2020.02.009
17. Guðjónsdóttir J, Paalsson KE, Sveinsdóttir GP (2021). Are the target exposure index and deviation index used efficiently? Radiography (Lond). Aug;27(3):903–907. doi: https://doi.org/10.1016/j.radi.2021.02.012. Epub 2021 Mar 9. PMID: 33707050.
18. Park H, Yoon Y, Kim J, Kim J, Jeong H, Tanaka N, Morishita J (2020). Use of clinical exposure index and deviation index based on national diagnostic reference level as dose optimization tools for general radiography in Korea. Radiat Prot Dosimetry; Nov 17: ncaa185. doi: https://doi.org/10.1093/rpd/ncaa185. Epub ahead of print. PMID: 33201240.
19. Seeram E (2004). Digital image processing. Radiol Technol. Jul–Aug;75 (6):435–52; 453–5. PMID: 15352557.
20. Seeram E, Seeram D (2008). Image Postprocessing in Digital Radiology-A Primer for Technologists. J Med Imaging Radiat Sci;39 (1):23–41. doi: https://doi.org/10.1016/j.jmir.2008.01.004. Epub 2008 Mar 22. PMID: 31051771.
21. Gonzalez RC, Woods RE (2018). Digital image processing. 4th ed. Toronto, Ontario: Pearson.
22. Baxes GA (1994). Digital image processing: principles and applications. New York: John Wiley & Sons, Inc.
23. Precht H, Gerke O, Rosendahl K, Tingberg A, Waaler D (2012). Digital radiography: optimization of image quality and dose using multi-frequency software. Pediatr Radiol; 42(9): 1112–8. doi: https://doi.org/10.1007/s00247-012-2385-3. Epub 2012 Apr 17. PubMed
24. American Association of Physicists in Medicine (2006). Acceptance testing and quality control of photostimulable storage phosphor imaging systems. Report No 93. College Park, MD
25. Lee W, Lee S, Chong S, Lee K, Lee J, Choi JC, et al (2020). Radiation dose reduction and improvement of image quality in digital chest radiography by new spatial noise reduction algorithm. PLoS ONE 15 (2): e0228609. https://doi.org/10.1371/journal.pone.0228609 PubMed
26. Feghali JA, Chambers G, Delépierre J, Chapeliere S, Mannes I, Adamsbaum C (2021) New image quality and dose reduction technique for pediatric digital radiography. Diagn Interv

Imaging;102(7–8):463–470. doi: https://doi.org/10.1016/j.diii.2021.01.009. Epub 2021 Feb 3. PubMed

27. Oh S & Kim JH, Yoo SY, Jeon T, Kim Y. (2021). Evaluation of the image quality and dose reduction in digital radiography with an advanced spatial noise reduction algorithm in pediatric patients. European Radiology. https://doi.org/10.1007/s00330-021-07942-6. PubMed
28. Damilakis J, Vassileva J (2021). The growing potential of diagnostic reference levels as a dynamic tool for dose optimization, Physica Medica, Volume 84, Pages 285–287, https://doi.org/10.1016/j.ejmp.2021.03.018
29. Tsapaki V (2020). Radiation dose optimization in diagnostic and interventional radiology: Current issues and future perspectives, Physica Medica, Volume 79, Pages 16–21, https://doi.org/10.1016/j.ejmp.2020.09.015
30. ICRP Publication 135 (2017). Diagnostic Reference Levels in Medical Imaging. Annals of the ICRP Volume 46, Issue 1, Pages 1–144
31. Seeram E, Brennan P (2017). Radiation Protection in Diagnostic X-Ray Imaging, Burlington, MA., Jones and Bartlett Learning.
32. American College of Radiology ACR-AAPM-SPR (2018). Practice parameter for diagnostic reference levels and achievable doses in medical x-ray imaging. https://www.acr.org/–/media/ACR/Files/Practice-Parameters/Diag-Ref-Levels.pdf (accessed 30 August 2020)

Chapter 4
Computed Tomography: A Technical Review

Keywords Computed tomography · Physical principles · Data acquisition · Image reconstruction algorithms · Iterative reconstruction algorithms · Deep learning image reconstruction algorithms in CT · Image display · Image storage · Attenuation · CT numbers · Multislice CT principles · Multislice CT detectors · Photon counting detector · Selectable scan parameters · Pitch · Radiation protection

4.1 Introduction

The first clinically useful Computed tomography (CT) scanner was invented by Godfrey Newbold Hounsfield in England. Later Hounsfield and Alan Cormack working in South Africa shared the 1979 Nobel Prize in Medicine or Physiology [1]. CT has better contrast resolution compared to digital radiography since it uses a computer to process attenuation data acquired from the patient (*data acquisition*). These attenuation values (μs) are subsequently converted into integers (0, a positive number, a negative number) referred to as CT numbers [1, 2–4] and all values are normalized to the attenuation of water (μ_{water}). Attenuation values are captured by special electronic detectors which convert electrical signals to digital data for processing. The computed uses a sophisticated mathematical process referred to as *image reconstruction* [5], which uses specialized algorithms to build up and display images of a patient's internal anatomy for diagnostic interpretation. CT images are essentially cross-sectional images of slices that are perpendicular to the long axis of the patient [6, 7]. Today, current CT scanners are referred to as Multislice CT (MSCT) scanners, since they are based on acquiring volume of tissue rather than one slice at a time.

The increasing use of CT has led to widespread concern about high patient radiation doses and current radiation protection in CT has focused on *dose optimization*, that is how to reduce patient dose and operate within the as low as reasonably

E. Seeram, *Dose Optimization in Digital Radiography and Computed Tomography*,
https://doi.org/10.1007/978-3-031-22871-1_4

achievable (ALARA) principle. Dose optimization will be described in detail in Chap. 5.

The purpose of this chapter is to describe the essential technical system components of CT, including data acquisition and image reconstruction, principles, as well as image post processing and image display considerations. Furthermore, the basics of MSCT scanners will be reviewed briefly. For a more detailed description of CT principles and technology, readers should refer to a comprehensive textbook on CT [1], and Chapters dedicated to CT principles and equipment, in other textbooks [2–4].

4.2 The CT Scanner: Fundamental Physics and Major System Components

The major system components of a CT imaging are illustrated in Fig. 4.1. These consist of data acquisition; image reconstruction; and image display, storage, and communication.

While data acquisition collects x-ray transmission readings by scanning the patient; image reconstruction uses these readings to build up a CT image of the patient's scanned anatomy. The transmitted radiation consists of attenuation values (μs) which are subsequently converted by the detector electronics into digital data for processing by a computer. This section will first explain the fundamental physics of attenuation, followed by a description of the major system components of data acquisition.

4.2.1 Attenuation Physics: An Essential Overview

Radiation attenuation is defined as the reduction of the intensity of a beam of radiation as it passes through matter or the patient's body. Attenuation behaves differently for a homogeneous beam (all photons in the beam have the same energy) and for a heterogeneous beam (all photons have different energies). The original experiments in CT performed by Hounsfield used a homogeneous beam from Americium (a gamma radiation source) because this beam holds true for a law of attenuation referred to as Beer-Lambert Law [3, 4, 6], described by the equation:

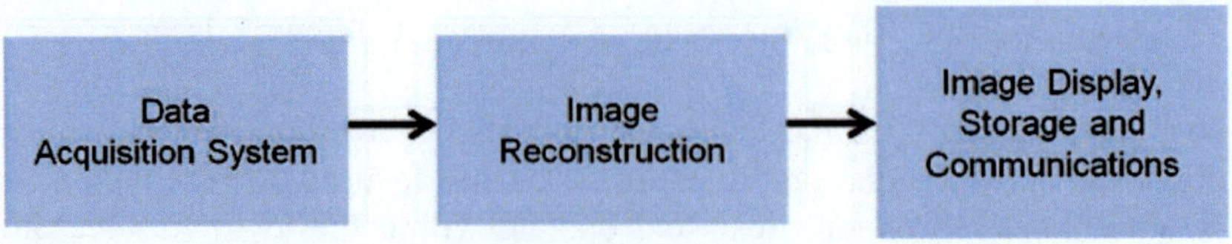

Fig. 4.1 The major system components of a CT imaging consist of data acquisition; image reconstruction; and image display, storage, and communication

$$I = I_0 e^{-\mu x}$$

where I is the transmitted intensity; I_0 is the original intensity; e is Euler constant (2.718); μ is the linear attenuation coefficient, or the fractional reduction in the intensity of a beam of radiation per unit thickness of the medium traversed; and x is the thickness of the object. The mathematical problem in CT is to calculate the linear attenuation coefficient, expressed as μ, which indicates the amount of attenuation that has occurred. The procedure to solve for μ is as follows:

$$I = I_0 e^{-\mu x}$$

$$\ln I_0 / I = {}^{-\mu x}$$

$$\mu = (I / x) \cdot (\ln I_0 / I)$$

where ln is the natural logarithm. In CT, however, the beam is attenuated by a given amount of tissue with a specific thickness Δx. Hence, the attenuation is expressed as:

$$I = I_0 e^{-\mu \Delta x}$$

4.2.2 *Attenuation and CT Numbers*

Attenuation values are converted into integers (0, a positive number, a negative number) referred to as *CT numbers* [1–4]. The system normalizes all tissue attenuation values to the attenuation of water (μ_{water}). CT numbers are computed using the following algebraic expression:

$$\text{CTNumber} = \frac{\mu_{\text{tissue}} - \mu_{\text{water}}}{\mu_{\text{water}}} \cdot K$$

where K is the scaling factor (contrast factor) of the CT manufacturer. In general, K is equal to 1000. CT numbers always are computed with reference to the attenuation of water. The CT number for water is 0, while it is +1000 for bone and −1000 for air on the Hounsfield scale. Furthermore, CT number ranges for bone, muscle, white matter, gray matter, blood, tumors, water, fat, lungs, and air, are 800–3000; 35–50; 36–46; 20–40; 13–18; 5–35; 0; −100; −150–400; and − 1000 respectively.

The overall process of obtaining CT numbers and converting them into a gray scale image is illustrated in Fig. 4.2, where higher CT numbers are assigned white, lower numbers black, and gray shades between black and white. This assignment is related to the attenuation characteristics of tissues. Bone attenuates more radiation and therefore is assigned white (the bone's appearance is the same on a film-screen

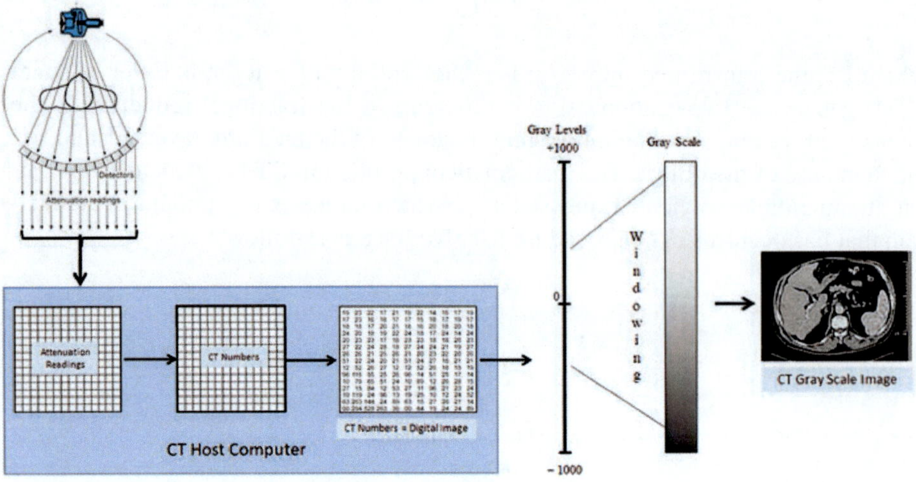

Fig. 4.2 The overall process of obtaining CT numbers and converting them into a gray scale image. (See text for further explanation)

image as it is on a digital image). Air attenuates very little radiation and appears black on film-screen and digital images. Digital image processing is applied to these integers through a process referred to as *windowing* to change the brightness and contrast of the displayed image. The range of CT numbers is defined as the window width and the center of the range is defined as the window level. Finally, the radiologist can manipulate the window width to alter image contrast, and the window level to alter image brightness.

4.2.3 Data Acquisition: System Components and Principles

The data acquisition components of the CT scanner are illustrated in Fig. 4.3. These components are housed in what is referred to as the *CT Gantry*, and include: an x-ray tube, a beam shaping filter called a *bow-tie filter*, pre-patient collimators. The x-ray tube is coupled to an array of detectors which collect radiation transmitted through the patient. The detectors send electrical signals to analog-to-digital converters (ADCs) which convert analog signals into digital data for processing by the CT host computer.

Data acquisition is a systematic collection of attenuation data from the patient through various scanning methods, which are designed to scan one slice per one rotation of the tube and detectors (single-slice acquisition) to multiple slices per single rotation of the x-ray tube and detectors. The latter scheme is referred to multi-slice CT scanning (MSCT) scanners which use a fan-beam geometry to scan a volume of tissue. The term geometry, or data acquisition geometry, refers to the size, shape, motion, and path traced by the x-ray beam. In MSCT scanning, the path

Fig. 4.3 The data acquisition components of the CT scanner. (See text for further explanation)

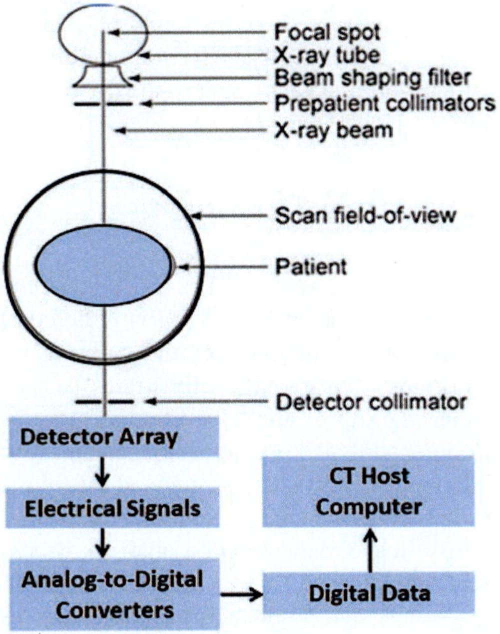

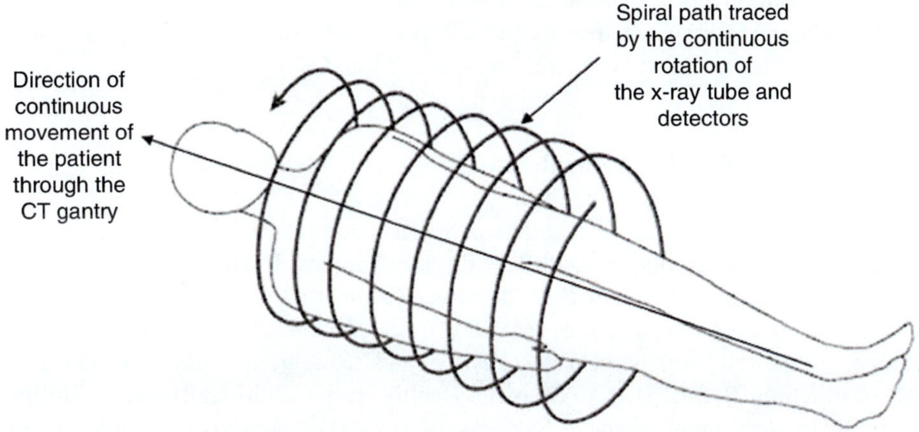

Fig. 4.4 In MSCT scanning, the path traced by the x-ray beam as the patient moves through the gantry aperture during scanning and is called a spiral or helical path

traced by the x-ray beam as the patient moves through the gantry aperture during scanning and is called a spiral or helical path (Fig. 4.4) MSCT scanners have evolved scanning 4 to 640 slices per revolution of the x-ray tube and detectors.

The purpose of the beam-shaping filter (bow-tie filter) as shown in Fig. 4.3, is to make the beam more uniform (appear to be homogeneous) at the detector and thus satisfy the Beer-Lambert law for calculating the linear attenuation coefficients. Additionally, collimation directs the beam through the slice of interest. The detectors measure the transmitted photons, and convert them to electrical signals which are subsequently converted into digital data by the analog-to-digital converters. The digital data are sent to the CT host computer for image reconstruction.

4.2.4 Image Reconstruction

Image reconstruction in CT is based on sophisticated mathematics and has evolved from simple back projection (BP) to filtered back projection (FBP) to more complex iterative reconstruction (IR) and currently artificial intelligence-based (AI-based) algorithms. This evolution illustrates basically that the problems associated with former algorithms are solved by the more current state-of-the-art algorithms. For example, while BP produced images that were blurred and were not diagnostic, the FBP image reconstruction removed this blurring using digital image processing filtering such as the convolution filtering [8–10].

FBP image reconstruction algorithms became the workhorse for decades, however, with the need for *low dose CT imaging*, the FBP convolution filters increased image noise and produced image artifacts. These problems as well as others (such as failure of the FBP algorithm to model the CT imaging system accurately [8]) provided the motivation to develop new algorithms that can reduce image noise while maintaining image sharpness and contrast, especially in low-dose CT imaging. IR algorithms were subsequently introduced into CT.

Iterative Reconstruction Algorithms

IR algorithms involve complex mathematics that model the CT imaging system accurately in order to operate with low dose techniques. Examples of these algorithms include Adaptive Statistical Iterative Reconstruction (ASiR) and Model-Based Iterative Reconstruction (MBIR) from GE Healthcare [10]; Iterative Reconstruction in Image Space (IRIS) and Sinogram Affirmed Iterative Reconstruction (SAFIRE) from Siemens Healthineers. Additionally, while Philips offers iDose4, Canon Medical Systems offers AIDR Adaptive Iterative Dose Reduction, along with AIDR 3-D Adaptive Iterative Dose Reduction [10].

The flowchart shown in Fig. 4.5, illustrates the fundamental steps of a typical IR algorithm. First, *measured projection data* sets are reconstructed with the FBP algorithm, to produce an initial CT image. This image is forward projected (mathematics) to produce *simulated data*. Next, the measure projection data is compared with the simulated data to create an updated image, which must meet a preset image quality criterion established by the user and is a part of the algorithm. If this criterion is not met, the iteration process continues over and over, until the difference between the measured and simulated data sets is very small. The iterative process stops when the final CT image matches the pre-set image quality criterion.

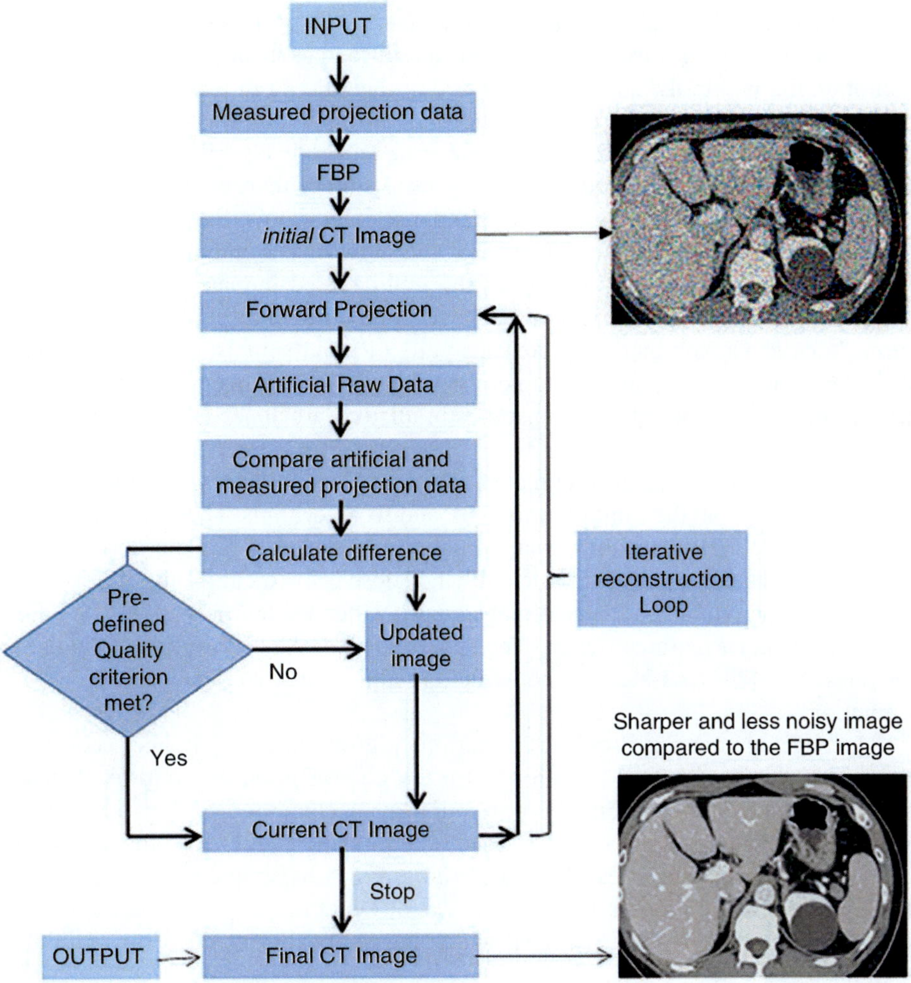

Fig. 4.5 Flow chart showing the primary steps of a typical iterative reconstruction algorithm, without modelling

Artificial Intelligence in CT Image Reconstruction

The literature is replete with articles on *Artificial Intelligence* (AI) describing what AI is; its roots; and what are its applications [11]. In this subsection, AI will be introduced briefly through a definition, and its application in CT image reconstruction. This is necessary in order to relate the use of AI-based CT image reconstruction to dose optimization. It is important for the reader interested in a further exploration of AI in general and AI in medical imaging, to refer to the relevant literature for more details [11–19].

There are several authoritative *definitions of AI*, however, it is beyond the scope of this book to address all of these. Cholet [17], defines AI as the "effort to automate intellectual tasks typically performed by humans". These tasks are performed by computers using programs (software) that can sense, reason, act, and adapt. Such as definition fits within the scope of other definitions offered in the literature as well [18–24]. While machine learning is a subset of artificial intelligence, deep learning is a subset of machine learning [11]. In *machine learning*, algorithms are trained to use patterns learned from data to perform tasks, rather than by explicit programming [12]. *Deep learning* on the other hand, uses algorithms that are characterized by the use of multilayered neural networks; this network is a unique type of artificial neural network (ANN) that resembles the multilayered human cognition system [13]. ANN is comprised of hundreds of basic computing units, and a learning algorithm, such as backpropagation, presents pairs of input signals and desired output decisions to train the weights (parameters) of the network. This mimics the process where the brain uses external sensory stimuli to learn to accomplish particular tasks [13].

The motivation for the development and use of AI techniques in CT image reconstruction stems from the limitations of FBP and IR algorithms. These limitations are as follows: the FBP algorithm, image quality is adequate when the dose is high; however, image noise and artifacts result when dose is lowered; IR algorithms solve this problem, producing adequate image quality when using *low-dose CT* (LD-CT) however, the noise texture often has been reported to appear blotchy, plastic-looking, or unnatural [25]. AI-based reconstruction algorithms attempt to solve these limitations.

The development and use of deep learning algorithms in CT image reconstruction is gaining widespread attention, with the goal of providing improved image quality, dose performance, and reconstruction speed compared with iterative reconstruction methods. It is not within the scope of this book, and hence this chapter to describe the basic principles of deep learning. For a basic understanding of deep leaning, the reader should refer to the works of Do et al. [21] and Chartrand et al. [22] as well as one by Seeram [23], which is specifically intended for radiologic technologists.

Deep Learning CT Image Reconstruction Algorithms

There are several examples of the use of deep learning image reconstruction (DLIR) in CT. For example, two such DLIR algorithms were approved by the U.S. Food and Drug Administration in 2019 namely; the Advanced Intelligent Clear-IQ Engine (AiCE) from Canon Medical Systems [24] and TrueFidelity™ from General Electric (GE) Healthcare [19]. It is not within the scope of this chapter to describe the details of each of these two DLIR algorithms and the reader is encouraged to refer to other works [19, 24–26].

A generalized framework for the AiCE and TrueFidelity™ consists of 3 steps (Fig. 4.6): deep learning algorithm development, training and optimization of the algorithm, and performance or verification of the algorithm. While TrueFidelity™ training and optimization involve the use of low-dose (LD) and high-dose CT data

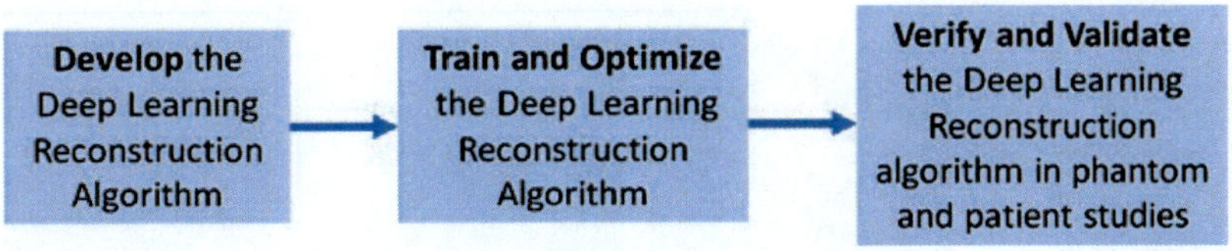

Fig. 4.6 A generalized framework for the AiCE and TrueFidelity™ deep learning image reconstruction consists of 3 steps: deep learning algorithm development, training and optimization of the algorithm, and performance or verification of the algorithm. (See text for further explanation)

reconstructed with the FBP algorithm; the AiCE uses iterative reconstruction (with modeling), and the data are fed into the deep learning reconstruction engine which performs the required operations to produce output images with high signal-to-noise ratio. During the training process, the output images are compared with a reference image using various parameters, including image noise, noise texture, low contrast resolution, and low contrast detectability. Differences are reported by the output image to the network via special DL network actions, which strengthens some equations and weakens others, and tries again. This process repeats until the output image is an accurate representation of the ground truth image. In deep learning, the term "ground truth" usually refers to correct labels prepared by experts [19]. During the final step, performance or verification, the algorithm is required to reconstruct clinical and phantom cases it has never encountered, including rare cases [19].

DLIR algorithms have been evaluated using phantom images as well a patient images, on noise reduction, noise texture, contrast-to-noise ratio, and low contrast detectability. The results show that these AI-based algorithms outperform iterative reconstruction algorithms on these parameters as well as fast reconstruction speed to meet the needs of clinical CT examinations [19, 24, 25, 27–29]. An example of a comparison between an IR and DL reconstruction algorithm on image noise is shown in Fig. 4.7.

4.2.5 Image Display/Storage/Communication

Image display/storage and communications represent the third system component of the CT scanner. Images are displayed on a monitor for viewing and interpretation, and can be subject to post processing to suit the viewing needs of the observer. Additionally, images are sent to the Picture Archiving and Communications System (PACS) for storage and electronic communications. PACS is an organized system that operates on computer networking and electronic communication infrastructure. A major element of such infrastructure is the use of standards to communicate images from the PACS to other remote monitors. This standard is the Digital Imaging and Communication in Medicine (DICOM) standard. For further details of DICOM the reader should refer to Bushberg et al. [4].

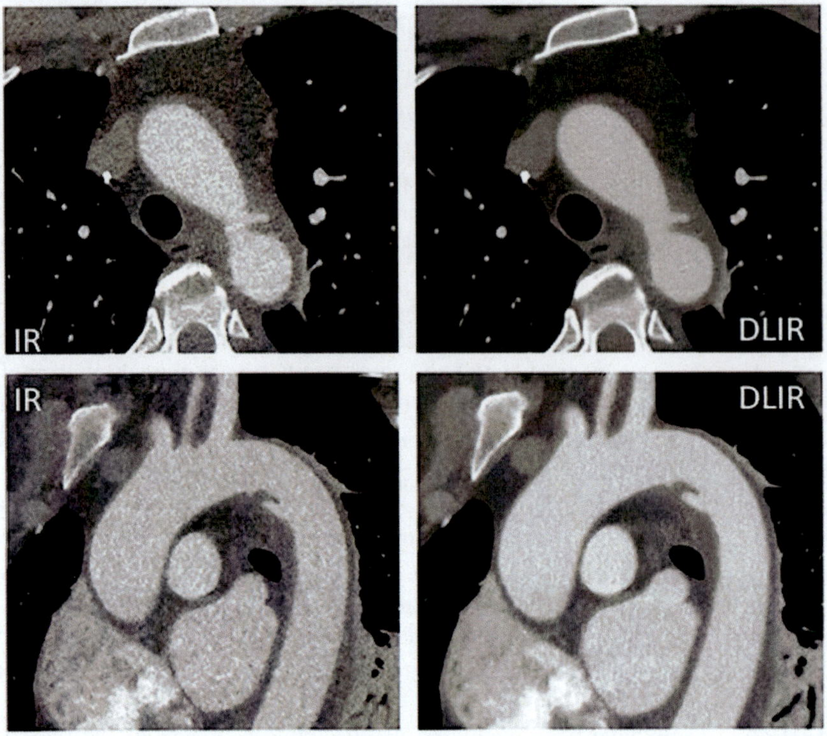

Fig. 4.7 A visual comparison of IR and DLIR algorithms on image noise. From Heinrich A, Streckenbach F, Beller E, Groß J, Weber MA, Meinel FG (2021). Deep Learning-Based Image Reconstruction for CT Angiography of the Aorta. Diagnostics (Basel). Nov 3;11 (11):2037. (Reproduced under the Creative Commons Attribution License which permits unrestricted use, distribution, and reproduction in any medium, provided the original work is properly cited)

4.3 Multi-slice CT Principles at a Glance

Multislice CT (MSCT) is state-of-the-art CT imaging system. MSCT is based on continuous data acquisition while the patient moves through the CT gantry aperture to collect a volume data set (multiple slices during a single rotation of the x-ray tube and detectors) as opposed to single-slice CT (SSCT) which acquires a one slice during a single rotation of the tube and detectors, as illustrated in Fig. 4.8. This strategy requires the use of not only specialized equipment (slip rings and x-ray tubes with higher output) but also two-dimensional (2-D) detector arrays, and continuous table movement during scanning. These technical principles will not be elaborated in this book, and therefore the reader should refer to the works of Seeram [1], Bushong [2], Wolbarst et al. [3], and Bushberg et al. [4] for a more detailed description of fundamental physical principles.

For the purposes of this chapter, it is noteworthy to review briefly MSCT detectors and selectable scan parameters since they play a significant role in radiation dose to the patient. CT dosimetry and factors affecting CT dose will be described in detail in Chap. 5, on CT Dose Optimization.

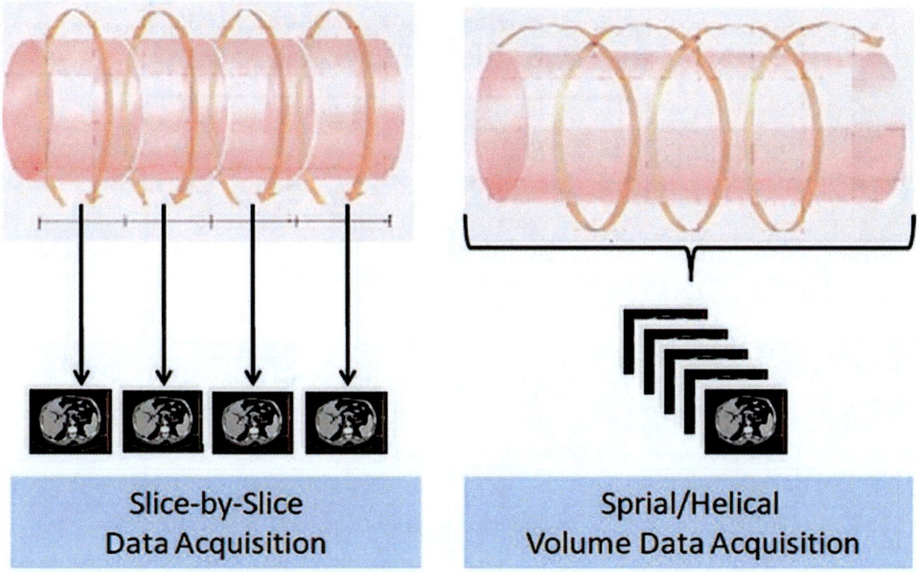

Fig. 4.8 MSCT is based on continuous data acquisition while the patient moves through the CT gantry aperture to collect a volume data set (multiple slices during a single rotation of the x-ray tube and detectors) as opposed to single-slice CT (SSCT) which acquires a one slice during a single rotation of the tube and detectors

4.3.1 MSCT Detectors: An Overview

The purpose of the detectors in CT is to capture x-ray photons transmitted through the patient and convert them into electrical signals and subsequently digital data for input into the CT host computer, which uses image reconstruction algorithms to create the CT image.

Scintillation Detectors and Photon Counting Detectors

MSCT detectors fall into two categories, namely; *scintillation detectors* and *photon counting detectors.* Furthermore, there are two types of scintillation detectors; energy-integrating detectors and dual-layer detectors as illustrated in Fig. 4.9. Scintillation crystals (such as cadmium tungstate; ceramic material made of high-purity, rare-earth oxides based on doped rare-earth compounds such as yttria; and gadolinium oxysulfide ultrafast ceramic) convert x-ray photons to light photons, which then are converted to electrical signals by photodiodes. Detector electronics called application-specific integrated circuits (ASIC) digitize the signals [30].

Furthermore, Philips Healthcare uses zinc selenide activated with tellurium in their dual-layer scintillator detectors [30].

Photon-counting detectors are now used in CT scanners. The US Food and Drug Administration (FDA) cleared the world's first photon-counting computed

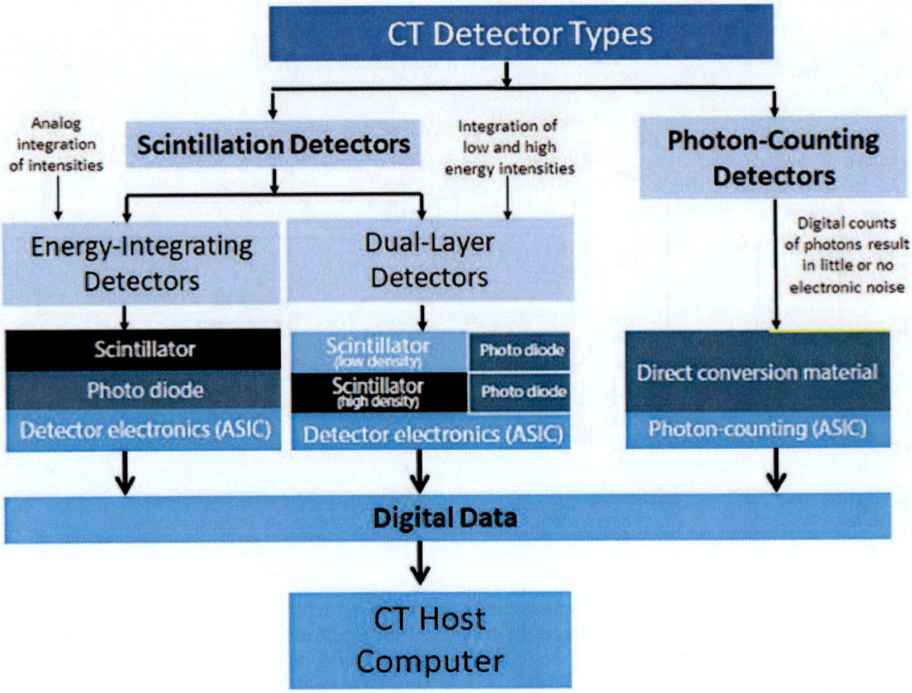

Fig. 4.9 MSCT detectors fall into two categories, namely; *scintillation detectors* and *photon counting detectors*. (See text for further explanation)

tomography (CT) scanner, the Siemens Naeotom Alpha, in 2021. These detectors use semiconductors such as cadmium telluride (CdTe) and cadmium zinc telluride (CZT) [18] because they can convert x-ray photons directly into electron hole pairs (electric charge). Studies which compared photon-counting detectors performance with energy-integrating detectors for example shat that "…. CT images obtained on a photon-counting detector CT scanner were rated as having superior spatial resolution and better critical structure visualization than those obtained on a conventional energy-integrating detector scanner, even with a substantial dose reduction" [31, 32].

MSCT Detector Designs

MSCT scanners use *2-D detector arrays* (Fig. 4.10) that acquire several slices per single rotation of the x-ray tube and detectors during volume scanning, as compared with 1-D detector arrays which acquire 1 slice per single rotation of the x-ray tube and detectors. The detector design shown in Fig. 4.10 shows 8 detector rows, which means that this detector will acquire 8 slices per rotation. The beam geometry needed to cover this 2-D detector is referred to as a cone beam. Slice selection with 2-D detectors is based on the detector configuration used, and have been discussed in the literature in detail [1, 2, 4],

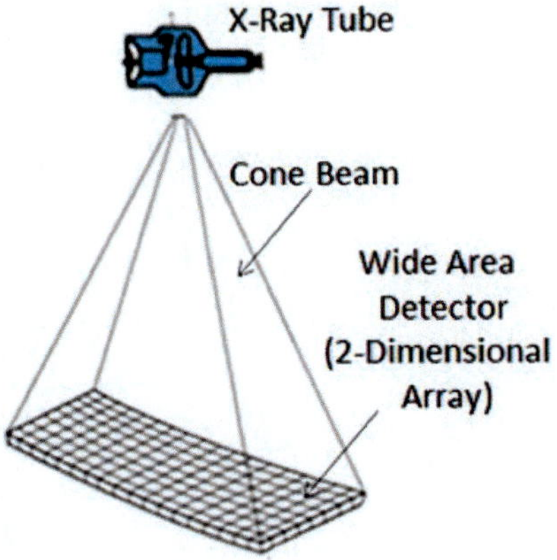

Fig. 4.10 MSCT scanners use *2-D detector arrays* that acquire several slices per single rotation of the x-ray tube and detectors during volume scanning, as compared with 1-D detector arrays which acquire 1 slice per single rotation of the x-ray tube and detectors. The detector in this figure shows 8 detector rows, which means that this detector will acquire 8 slices per rotation

4.3.2 Selectable Scan Parameters

There are several parameters affecting the dose in MSCT imaging, however, the technologist is concerned with what has been referred to as *selectable scan parameters*. These include the scan mode, exposure factors (kV, mA, and scan time), gantry rotation time, pitch, scan length, collimation, and slice width. In particular, the pitch determines image quality needed and is related to patient dose. The International Electrotechnical Commission (IEC) [33] provides a universal definition of pitch for MSCT scanner technology as: the pitch (P) is equal to the distance the table travels per rotation (d)/total collimation (W). The total collimation is equal to the number of slices (M) times the collimated slice thickness (S). Algebraically, the pitch is expressed as:

$$P = d / W \text{ or } P = d / M \times S.$$

In addition, the pitch is related to dose and image quality. As the pitch increases from 1 to 2, image quality is compromised (image becomes noisy), as shown illustrated in Fig. 4.11. The relationship between dose and pitch will be described in Chap. 5 on dose optimization.

4.4 Radiation Protection

The need for radiation protection in CT is an important consideration based on the fact that technological advances in CT have resulted in its increasing use in practice. This fact provided the motivation to explore the impact of these changes on patient dose. A number of studies show that patient doses in CT is higher relative to other

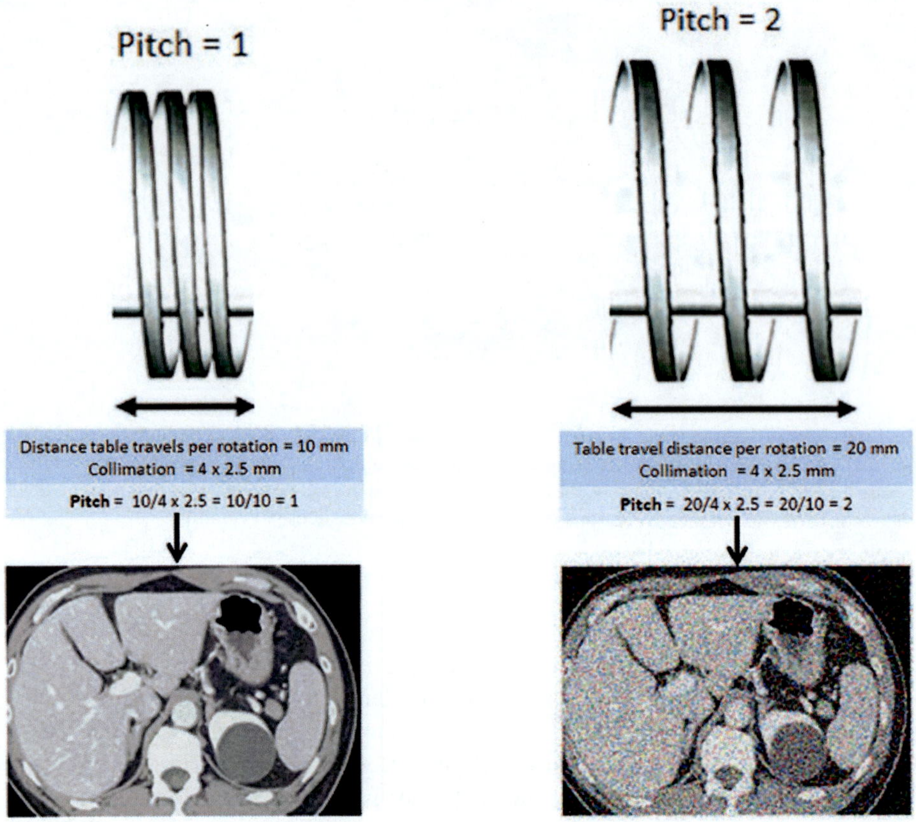

Fig. 4.11 As the pitch increases from 1 to 2, image quality is compromised (image becomes noisy). Pitch is also related to dose to the patient (Chap. 5)

imaging modalities [34–36], and therefore the literature stresses the need to optimize dose in CT, thus providing the best possible radiation protection strategies for the patient. Dose optimization in CT will be described in detail in Chap. 5.

References

1. Seeram E (2022). Computed Tomography: Physical Principles, Patient Care, Clinical Applications, and Quality Control. 5th Edition, Philadelphia, PA: Elsevier Inc.
2. Bushong S (2021). Radiologic Science for Technologists. 12th ed. St Louis, MO: Mosby-Elsevier.
3. Wolbarst AB, Capasso P, Wyant AR (2013). Medical Imaging: Essentials for Physicians. Hoboken, NJ: Wiley-Blackwell.
4. Bushberg JT, Seibert JA, Leidholdt EM, Boone JM (2021). The Essential Physics of Medical Imaging. 4th ed. Philadelphia, PA: Lippincott Williams & Wilkins.
5. Herman GT (1980). Image Reconstruction from Projections. New York, Academic Press.

6. Hounsfield GN (1973). Computerized transverse axial scanning (tomography): Part 1. Description of system. Br J Radiol; 46:552, 1016–1022 https://doi.org/10.1259/0007-1285-46-552-1016
7. Hounsfield, G. N. (1976). Picture quality of computed tomography. American journal of Roentgenology, 127(1), 3–9.
8. Hsieh J (2008). Adaptive statistical iterative reconstruction. Whitepaper, GE Healthcare
9. Seibert JA (2014). Iterative reconstruction: how it works, how to apply it. Pediatr Radiol; 44(suppl 3):431–439. doi:https://doi.org/10.1007/s00247-014-3102-1.
10. Qiu D, Seeram E (2016). Does iterative reconstruction improve image quality and reduce dose in computed tomography? Radiol Open J; 1(2):42–54. doi:https://doi.org/10.17140/ROJ-1-108.
11. Boden MA (2018). Artificial Intelligence: A Very Short Introduction. Oxford University Press; 2018.
12. Erickson BJ, Korfiatis P, Akkus Z, Kline TL (2017). Machine learning for medical imaging. Radiogr; 37(2):505–515. doi:https://doi.org/10.1148/rg.2017160130
13. Lee J-G, Jun S, Cho Y-W, et al (2017). Deep learning in medical imaging: general overview. Korean J Radiol;18(4):570–584. doi:https://doi.org/10.3348/kjr.2017.18.4.570
14. Chollet F (2018). Deep Learning with Python. Manning Publications Co.
15. Ertel W (2017). Introduction to Artificial Intelligence. 2nd ed. Springer International Publishing.
16. Flasinski M (2017). Introduction to Artificial Intelligence. Springer International Publishing.
17. Wooldridge M (2018). Artificial Intelligence. Ladybird Books Ltd.
18. Wagner JB (2019). Artificial intelligence in medical imaging. Radiol Technol; 90(5):489–501.
19. Hsieh J, Liu E, Nett B, Tang J, Thibault J-B, Sahney S (2019). A new era of image reconstruction: TrueFidelity. https://www.gehealthcare.ru/jssmedia/040dd213fa89463287155151fdb01922.pdf. Published 2019. Accessed January 2, 2020.
20. Geyer LL, Schoepf UJ, Meinel FG, et al. State of the art: iterative CT reconstruction techniques. Radiol. 2015;276(2):339–357. doi:https://doi.org/10.1148/radiol.2015132766
21. Do S, Song KD, Chung JW. Basics of Deep Learning: A radiologist's guide to understanding published radiology articles on deep learning. Korean J Radiol. 2020;21(1):33–41. doi:https://doi.org/10.3348/kjr.2019.0312
22. Chartrand G, Cheng PM, Vorontsov E, et al. Deep learning: a primer for radiologists. Radiogr. 2017;37(7):2113–2131. doi:https://doi.org/10.1148/rg.2017170077
23. Seeram E (2020). Computed Tomography Image Reconstruction. Radiol Technol. 2020 Nov; 92(2):155CT-169CT. PMID: 33203780.
24. Boedeker K (2019). AiCE deep learning reconstruction: bringing the power of ultra-high resolution ct to routine imaging. Canon Med Syst; 2:28-33
25. Zhang, M., Gu, S. & Shi, Y. (2022). The use of deep learning methods in low-dose computed tomography image reconstruction: a systematic review. Complex Intell. Syst. https://doi.org/10.1007/s40747-022-00724-7
26. Suzuki K. Overview of deep learning in medical imaging. Radiol Phys Technol. 2017;10(3):257–273. doi:https://doi.org/10.1007/s12194-017-0406-5
27. Shan H, Zhang Y, Yang Q, et al (2018). 3D convolutional encoder–decoder network for low-dose CT via transfer learning from a 2D trained network. IEEE Trans Med Imaging; 37(6):1522–1534. doi:https://doi.org/10.1109/TMI.2018.2832217
28. Shan H, Padole A, Homavounieh F, et al (2019) Competitive performance of a modularized deep neural network compared to commercial algorithms for low-dose CT image reconstruction. Nat Mach Intell; 1:269–276. doi:https://doi.org/10.1038/s42256-019-0057-9
29. Akagi M, Nakamura Y, Higaki T, et al (2019). Correction to: deep learning reconstruction improves image quality of abdominal ultra-high-resolution CT. Eur Radiol; 29(8):4526–4527. doi:https://doi.org/10.1007/s00330-019-06249-x
30. Shefer E, Altman A, Behling R, et al (2013). State of the art of CT detectors and sources: a literature review. Curr Radiol Rep; 1(1):76–91. doi:https://doi.org/10.1007/s40134-012-0006-4.

31. Zhou W, Lane JI, Carlson ML, Bruesewitz MR, Witte RJ, Koeller KK, Eckel LJ, Carter RE, McCollough CH, Leng S (2018). Comparison of a Photon-Counting-Detector CT with an Energy-Integrating-Detector CT for Temporal Bone Imaging: A Cadaveric Study. AJNR Am J Neuroradiol;39(9):1733–1738. doi: https://doi.org/10.3174/ajnr.A5768. Epub. PMID: 30093479; PMCID: PMC6128765.
32. Benson JC, Rajendran K, Lane JI, Diehn FE, Weber NM, Thorne JE, Larson NB, Fletcher JG, McCollough CH, Leng S. (2022) A New Frontier in Temporal Bone Imaging: Photon-Counting Detector CT Demonstrates Superior Visualization of Critical Anatomic Structures at Reduced Radiation Dose. AJNR Am J Neuroradiol;43(4):579–584. doi: https://doi.org/10.3174/ajnr.A7452. Epub 2022 Mar 24. PMID: 35332019; PMCID: PMC8993187.
33. IEC (1999). Medical electrical equipment-60601 Part 2-44: particular requirements for the safety of x-ray equipment for CT. Geneva, Switzerland: International Electrotechnical Commission.
34. Berrington de Gonzalez A, Mahesh M, Kim KP, et al (2009). Projected cancer risks from computed tomographic scans performed in the United States in 2007. Arch Intern Med; 169(22):2071–2077. doi:https://doi.org/10.1001/archinternmed.2009.440
35. Van der Molen AJ, Stoop P, Prokop M, Geleijns J (2013). A national survey on radiation dose in CT in The Netherlands. Insights Imaging; 4(3):383–390. doi:https://doi.org/10.1007/s13244-013-0253-9.
36. Mathews JD, Forsythe AV, Brady Z, et al (2013). Cancer risk in 680,000 people exposed to computed tomography scans in childhood or adolescence: data linkage study of 11 million Australians. BMJ; 346: f2360. doi:https://doi.org/10.1136/bmj.f2360.

Chapter 5
Dose Reduction and Optimization Strategies in Computed Tomography

Keywords CT dose reduction · CT dose optimization · Radiation risks · Stochastic effects · Deterministic effects · Cancer risks from CT imaging · Selectable scan parameters · Tube current modulation · kV optimization · Image reconstruction optimization · Deep learning image reconstruction · Diagnostic reference levels

5.1 Introduction

The introduction of CT imaging in 1970s in medicine and specifically in diagnostic imaging has provided numerous clinical benefits in the detection, evaluation and diagnosis of human diseases. Hricak et al. [1] have shown that the growth and technical advances have resulted in more effective surgical treatments and have resulted in eliminating the need for some invasive exploratory procedures. Furthermore, these advances have helped decrease inpatient hospital stays; improve cancer, stroke, cardiac, and trauma diagnosis and treatment; and enabled rapid diagnosis of life-threatening vascular conditions [1]. Additionally, the there has been an increase in the use of CT imaging in clinical medicine. For example, CT use in hospital emergency departments has increased significantly from 2005 through 2013 [2], in an effort to prioritize and reduce wait times in the care and management of patients [3].

This increase in the use of CT has prompted research on that patient doses from CT examinations, and the results have shown that these doses are high relative to other radiography examinations [4–6]. Additionally, it has been reported that CT doses typically range from 5 to 50 mGy to each organ within the image field [7]. As of 2011, CT contributed the highest collective amount of medical radiation exposure in the United States compared with any other medical imaging modality [1].

E. Seeram, *Dose Optimization in Digital Radiography and Computed Tomography*,
https://doi.org/10.1007/978-3-031-22871-1_5

5.2 Motivation for Dose Optimization in CT

The few studies cited in the introduction above, stress the need to optimize the doses in CT imaging. The general motivation of dose optimization in CT and digital radiography has been presented in Chap. 1 and include radiation risks which have been studied extensively. The kinds of biological effects observed have been summarized by Hendee and O'Connor [8], who drew attention to Radiation Effects Research Foundation (RERF) data and models based on Japanese atomic bomb survivors that form the basis of the Biological Effects of Ionizing Radiations VII (BEIR VII) report [8].

These effects of radiation fall into 2 broad categories: stochastic effects and deterministic effects [4, 9] and have been reviewed in Chap. 1. In summary, *stochastic effects* are effects for which the probability of occurrence increases with increasing dose, and for which there is *no threshold dose* (any amount of radiation, no matter how small, has the potential to cause harm). Stochastic effects also are called late effects, and examples of these effects include cancer and genetic damage [4, 9]. Stochastic effects are considered a risk from exposure to the low levels of radiation used in medical imaging, including CT examinations [4, 9]. *Deterministic effects* are those effects for which the severity of the effect (rather than the probability) increases with radiation dose and for which there is a threshold dose. Examples of deterministic effects include skin burns, hair loss, tissue damage, and organ dysfunction [4, 9].

An important and significant noteworthy point about these effects is that the RERF data show that at high doses (100 mSv and higher), the evidence of increased cancer is statistically significant [8]. In medical imaging using x-rays, the doses to patients are much lower, and therefore in order to understand the cancer risks, dose response models have been proposed to extrapolate cancer risks from a high-dose situation to the risks of the low doses used in diagnostic imaging. These models have been introduced in Chap. 1, and will not be described further in this Chapter. In summary, there are tow categories of models, namely; the linear dose-response models and nonlinear dose-response models [9]. Furthermore, there are two types of linear dose-response models: a *linear dose-response model without a threshold* (LNT) and a *linear dose-response model with a threshold.* While the former is used to show that no amount of radiation is considered safe and that any dose, no matter how small, carries some degree of risk, the latter proposes that no adverse effect from a dose below a certain level, known as the *threshold dose*, is observed. An effect occurs only when the threshold dose is reached [9].

The next obvious question is which model is used in diagnostic radiology. The answer to this question is provided Hendee and O'Connor [8]. They argue that the LNT model is more commonly used because it is a simpler and conservative approach, which is more likely to overestimate cancer risk at low doses than to underestimate risk of cancer induction. Furthermore, Hendee and O'Connor [8], stress that because the doses in diagnostic radiology are so low, there is no evidence that the LNT model is effective at estimating cancer induction risk [8]. This has led to what is referred to as "*the Pro-LNT/Anti-LNT Debate*" in a commentary presented by Pamela J. Sykes, Flinders Centre for Innovation in Cancer, College of Medicine and Public Health, Flinders University, Adelaide, South Australia, Australia [10]. The author points out that "Until those on one side of the debate can convince the other, it would be sensible to move forward toward a graded (risk-based) approach to regulation, where the stringency of control is commensurate with the risk, resulting hopefully in more sensible practical thresholds. This approach is gradually being put forward by international radiation protection advisory bodies" [10].

5.2.1 Literature on Cancer Risks from CT

The literature is replete with articles on cancer risks from CT, and studies have emerged as a result of the relatively high doses from CT examinations [4, 6, 12], particularly for pediatric patients. It is not within the scope of this Chapter to describe details of these studies, however, a few examples of the results of selected examples are noteworthy. For example, Pearce et al. [13] concluded that there is a risk of cancer from diagnostic x-ray exposures and that the risk is very low, however the risk is detectable, so statements that radiation risks associated with diagnostic exposures are simply undetectable must be questioned. In one of the first large studies to derive direct estimates of cancer risk from CT instead of extrapolating risk from high-dose exposures. Matthews et al. [7] found "an increased risk of cancer in the children and adolescents who had CT scans from 1985 through 2005" but also "also suggested that estimating risks from many studies has been difficult because of relatively small study sizes and selection bias".

Hall and Brenner [14] in an article entitled "Cancer risks from Diagnostic Radiology" published in the *British Journal of Radiology* point out that "most of the collective dose from diagnostic radiology comes from high-dose (in the radiological context) procedures such as CT, interventional radiology and barium enemas; for these procedures, the relevant organ doses are in the range for which there is now direct credible epidemiological evidence of an excess risk of cancer, without the need to extrapolate risks from higher doses. Even for high-dose radiological procedures, the risk to the individual patient is small, so that the benefit/risk balance is generally in the patients' favour". An interesting study by Shao et al. [15] showed that "CT scans may be associated with an increased risk of thyroid cancer and leukemia in adults and in those diagnosed with NHL at a younger age".

Other studies followed and reported in the literature. Examples of two such studies are those of Moghadama [16] published in 2021, and Zewdea et al. [17] published in 2022. While the purpose of the former study was "To estimate percentage of patients undergoing multiple CT exams leading to cumulative effective dose (CED) of more than 25, 50, 75 and 100 mSv in one year and assess per capita and the collective effective dose", the latter study was "to estimate cumulative organ doses and age- and gender-stratified cancer mortality risks in patients undergoing recurrent computed tomography (CT) exams." The conclusions are as follows:

- "The alarming high CED received by large number of patients and with high collective dose to population requires urgent actions by all stake holders in the best interest of patient radiation safety" [16]
- "Organ doses over 100 mGy for most organs and for some organs ≥200 mGy with unignorable associated lifetime attributable cancer mortality rates were found" [17]

A final study to be cited in this Chapter is one by Hong et al. [18] entitled "Association of Exposure to Diagnostic Low-Dose Ionizing Radiation with Risk of Cancer Among Youths in South Korea" and published in JAMA Network Open. This original investigation explored the following:

1. Cancer Risk Associated with CT Scanning
2. Risks of specific cancers
3. Cancer Risk According to Type of First Exposure, and the
4. Cancer Risk According to Number of Exposures

The investigators concluded that "the associations we found of diagnostic low-dose ionizing radiation with increased incidence of cancer in youths suggest that there is incentive to limit radiation doses to as low as reasonably achievable and to only scan when justified. Medical professionals should weigh the benefits of diagnostic low-dose ionizing radiation with the associated risks to justify each decision" [18].

5.2.2 Effects of Low-Dose Chest CT on Chromosomal DNA

An interesting and important study is one by Sakane et al. [19] where *low-dose CT* examinations were compared to *standard dose CT* of the chest in exploring "the effects on DNA double-strand breaks and chromosome aberrations (CAs) in peripheral blood lymphocytes" Sakane et al. [19] concluded that "no effect of low-dose CT on human DNA was detected. In the same setting, DNA double-strand breaks and chromosome aberrations increased after standard-dose CT".

5.2.3 In Summary

The above are examples of studies on CT doses that are defined as high, standard, and low on biological effects. These studies were cited to provide the motivation for dose optimization in CT. In this regard, it is important to repeat here, the conclusions drawn by Moghadama [16], which is "The alarming high CED (Cumulative Effective Dose) received by large number of patients and with high collective dose to population requires *urgent actions by all stake holders in the best interest of patient radiation safety*" [16]. These "urgent actions" form the basis for the content of the rest of this Chapter.

5.3 Radiation Protection Principles

Current standards of radiation protection include one of the fundamental guiding principles of the ALARA (As Low as Reasonably Achievable) philosophy of the International Commission on Radiological Protection (ICRP) [11]. This principle identifies the notion of dose reduction and dose optimization, which are intended to minimize stochastic effects and to prevent deterministic effects. The ICRP recommends approaches to dose optimization associated with radiography equipment and daily operations [20]. These were reviewed in Chap. 1 and involves at least four elements, namely; the imaging equipment, the adequacy of the imaging equipment, technical parameters of the examination, and diagnostic reference levels.

In CT, dose optimization requires that the user understands the elements of CT dosimetry, reviewed in the next section.

5.4 Elements of CT Dosimetry at a Glance

CT dosimetry refers to the measurement of the dose to the patient. As shown in Fig. 5.1, the typical dose distribution is a bell-shaped curve, given by the function D(z), where D is the dose and *z is the longitudinal axis* of the patient. D(z) is extremely important to the CT dose because this is the dose distribution, or dose profile, that is measured. CT dose measurements are characterized by at least three commonly used metrics (dose quantities): the Computed Tomography Dose Index (CTDI), the Dose Length Product (DLP) and the Effective Dose (ED). While the units of the CTDI and the DLP are expressed in milligrays (mGy), the ED is expressed in millisieverts (mSv). In CT dose optimization studies, these dose quantities are measured and reported. The ED relates the radiation exposure to risk and

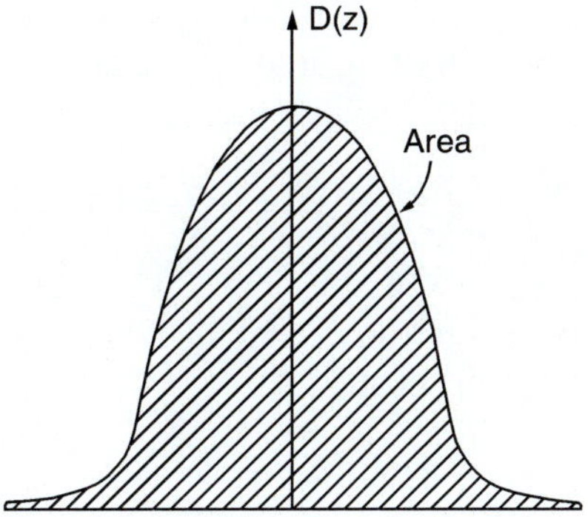

Fig. 5.1 The typical dose distribution in CT imaging is a bell-shaped curve, given by the function D(z), where D is the dose and *z is the longitudinal axis* of the patient. D(z) is extremely important to the CT dose because this is the dose distribution, or dose profile, that is measured

is considered the best method available to estimate stochastic radiation risk [1]. Still, effective dose is an estimate only, based on weighting factors applied to the body's tissue or the organ being irradiated [9].

As discussed by McNitt-Gray [21], CTDI dosimetry is the current international standard for estimating CT radiation dose. The CTDI has evolved along the technical advances made in CT during the years and readers interested in such evolution should refer to a paper by Seeram [22]. The introduction of spiral/helical CT beam geometry used for volume scanning led to the development of the $CTDI_{volune}$ ($CTDI_{vol}$) expressed algebraically as:

$$CTDI_{vol} = CTDI_{w} / \text{pitch}$$

where $CTDI_w$ is the weighted CTDI, intended to addresses the average dose in *the x-y axis of the patient*. When the pitch is 1, the $CTDI_{vol}$ equals the $CTDI_W$. A significant point to note is that the value of the $CTDI_{vol}$ is the same whether a 1-mm or 100-mm length of tissue of the patient is scanned. This problem is solved by the introduction of the DLP in order to provide a much more accurate value of the dose for a defined length of tissue. The DLP provides a measure of the total dose for a CT, and it is proportional to the scan length (L). Algebraically, the DLP is expressed as:

$$DLP = CTDI_{vol} \times L$$

For example, the data for 2 CT scans of the abdomen are: Abdomen 1-the scan length is 16 cm, and the DLP is 160 mGy-cm. For abdomen 2, the scan length is 32 cm, and the DLP is 320 mGy-cm, that is, if the scan length doubles, the DLP doubles.

CT dose reporting also include the ED in order to relate radiation exposure to risk, and it takes into account that different tissues have different radiosensitivities. Because only parts of the body (rather than the entire body) are exposed in CT imaging, the risk of stochastic effects is proportional to the effective dose rather than to the tissue dose [9]. The effective dose is expressed algebraically as:

$$ED = \Sigma W_T\, H_T$$

where H_T is equal to the organ or tissue dose, and W_T is the tissue weighting factor. Using DLP values, the

$$ED = k \times DLP$$

where k is a constant for different body parts. For more information on ED for different body parts and modalities, the reader should refer to McKnitt-Gray [21] and Wolbarst et al. [23].

5.5 Factors Affecting the Dose in CT: An Overview

Understanding dose optimization in CT requires that users (operators) have a firm grasp of note only the technical factors affecting the dose to the patient having CT examinations, but also all image quality factors as well since dose and image quality are closely related [24].

It is not within the scope of this Chapter to describe all image quality and dose factors in detail and therefore, readers should refer to the Bushberg et al. [9], Seeram [22], Wolbarst et al. [23] and Goo [24] for more detailed descriptions, however the following are noteworthy and are briefly reviewed in the next subsections.

5.5.1 Image Quality Considerations

There are at least four image quality parameters that are related significantly to dose. They parameters include image noise; contrast-to-noise resolution; spatial resolution; and artifacts. Only the first three will be considered for the purposes of this Chapter. According to Goo [24]:

- *Noise* is defined as a random variation of CT numbers, and is inversely proportional to square root of radiation dose; inversely proportional to tube voltage; inversely proportional to fourth power of spatial resolution; influenced by section thickness and reconstruction algorithm.
- *Contrast-to-noise resolution* is defined as the ability to distinguish between different CT numbers. Not significantly changed at different tube voltages in most materials except for some with high atomic numbers such as iodine (increased at low tube voltage); low contrast resolution greatly influenced by image noise level.
- *Spatial resolution* is defined as the ability to distinguish small details of object, and it is inversely proportional to focal spot size and detector collimation; influenced by reconstruction algorithm.

5.5.2 Significant Dose Parameters

Important CT dose parameters have been identified by Goo [24], and include kV, mAs, pitch, $CTDI_w$, $CTDI_{vol}$, DLP, and ED. The key factors that radiologic technologists can select to optimize dose are *exposure technique factors* (mAs and kV), *pitch, collimation and slices, Automatic Exposure Control* (AEC), *noise index, overbeaming* and *overranging*, and *noise-reduction image reconstruction algorithms.*

The key relationships that are noteworthy in CT dose optimization however, are:

- The dose is directly proportional to the mAs, that is doubling the mAs doubles the dose
- The dose is proportional to the square of the kilovolt peak (kV^2). This exponential expression poses some difficulty for technologists in practice because this power [2] can vary from 2.5 to 3.1 and depends on the patient's size [25]. An interesting study by Maldjian and Goldman [26], who showed that decreasing the kilovolt peak from 140 to 120 kVp reduces the dose by 28% to 40% for a typical phantom. The authors reported that further decreasing to 80 kVp reduces the dose by approximately 65%.
- Dose (D) is inversely related to the pitch (P). If the pitch increases by 2, the dose is reduced by 0.5 mGy. This relationship is an inverse one and is expressed:

$$D \alpha 1/P$$

- When using the AEC, the tube current increases when the pitch is increased, and therefore it is not best practice to reduce exposure with multislice CT scanners
- Generally, as the collimation width increases (wider beam = thicker section), the dose decreases. When decreasing the section thickness (T, the reconstructed or nominal thickness), the exposure must be increased to maintain the same signal-to-noise ratio (SNR) as a thick section. For example, a 2.5-mm section requires 2 times more exposure than a 5-mm section [24]. The relationship for the noise in the image is an inverse one expressed as follows:

$$\text{Noise}\, \alpha\, 1/T$$

- A technique referred to as *adaptive or dynamic collimation* is used to address *overranging* and *overbeaming* in CT in an effort to reduce the dose to the patient at the beginning and end of the scanning [27]. While overranging refers to the use of additional rotations before and after the planned length of tissue so the first and last images can be reconstructed, overbeaming is the excess dose beyond the edge of the detector rows per rotation of a multisection. Furthermore, Christner et al. [27] showed that dynamic collimation can reduce the dose by approximately 40%.
- The use of the AEC requires that a preselected *image quality index* also referred to as a *reference or target image quality*, be established before scanning the patient [24]. The index is stored in the CT scanner by the manufacturer before shipping and is an operator-selectable parameter [24]. As the index increases, the patient dose decreases but at the expense of a noisy image.
- Accurate *patient positioning* is an essential task of the technologist during CT scanning, If the patient is not centered in the CT gantry isocenter (not centered accurately) in the scan field-of-view, image noise and patient dose increase because of poor bowtie filter performance. As demonstrated by Toth et al. [28] miscentering the patient by as little as 3 cm increased surface dose by 18%, and miscentering by 6 cm increased surface dose by as much as 41%

- *Iterative reconstruction algorithms* and now used routinely in low-dose (LD) CT imaging. These algorithms are intended to reduce image noise and minimize the higher radiation dose [29–31]. A number of studies have demonstrated reductions in radiation dose using iterative reconstruction that vary from 30% to 50% [26, 30]. These studies have included dose reductions in pediatric studies, CT abdominal studies, and CT angiography.

5.6 Dose Optimization Strategies in CT

The idea of dose optimization was introduced in Chap. 1, and subsequently elaborated in Chap. 3 for digital radiography. In Chap. 1, the distinction between dose reduction and dose optimization was described. In review, operators working within the ALARA philosophy of the ICRP balance the need for patient radiation protection with the need for acquiring high-quality diagnostic images. In doing so they are guided by the meaning of two terms used in the literature, that is; *reduction* and *optimization*. While reduction means to "reduce or diminish in size, amount, extent, or number" the term optimization demands a more rigorous effort to be effective in adhering to the ALARA philosophy. Optimization means "an act, process, or methodology of marking something (as a design, system, or decision) as fully perfect, functional, or effective as possible" [32]. Perhaps the key difference between these two terms is that *dose optimization takes into primary consideration that the image quality must not be compromised as the dose is lowered.* This is where the rigor of an optimization study lies. For example, a CT *dose optimization study* would involve an extensive component related to the methodology used in the study. Such methodology must use reliable and valid tools and procedures for the dosimetry, image acquisition, and evaluation of image quality using human observers, keeping in mind the nature of the detection task. The details within this definition of optimization will be discussed further in Chap. 6.

CT dose reduction involves strategies to adjust and control the technical factors affecting the dose with the goal of decreasing patient dose [33]. The effect of these factors (eg, mAs, kV, pitch, scan field-of-view, beam collimation, AEC, overbeaming and overranging, and iterative image reconstruction) on patient dose has been researched by multiple authors [22–29, 34].

CT dose optimization strategies have been explored and the results published in the literature. A few illustrative examples include:

1. Sohaib et al. [35] examined the effect of reducing milliampere seconds on image quality and patient dose in sinus CT examinations. The authors reported a dose reduction from 13.5 mGy at 200 mAs to 3.1 mGy at 50 mAs (P = .05) without loss of image quality. The study used an observer performance method that involved visual grading analysis.
2. Russell et al. [36] examined dose-image quality in neck volume CT and showed that automatic tube current modulation reduced the CTDI by 20% with the noise

index set at 11.4 and by 34% with the noise index set at 20.2. The dose reductions were made without compromising image quality significantly.

5.7 Other Useful Tools for Dose Reduction and Optimization in CT

There are several additional tools for dose reduction and optimization in CT and these include the use of *Diagnostic Reference levels (DRLs)*, *artificial intelligence-based (AI-Based) image reconstruction,* and more recently the use of *photon counting detectors (PCD)* in CT imaging.

5.7.1 The Use of DRLs

The ICRP and other Radiation Protection Authorities such as the National Council on Radiation Protection and Measurements (NCRP) have established dose limits not only for occupationally exposed individuals such as radiologists and technologists, pregnant workers, but for members of the public as well. There are no dose limits however for patients having x-ray imaging procedures including diagnostic radiographic and fluoroscopic as well as CT examinations, for the management of their medical illness. This scenario was addressed by the ICRP, as early as 1990, and in this regard the ICRP [37] introduced the notion of *Diagnostic Reference Levels* (DRLs), and recommends the use of DRLs for patients. Dose limits and DRLs are not the same thing. A thorough review of DRLs for technologist is one by Seeram and Brennan [38] published in Radiologic Technology. Furthermore, a brief summary of the essential elements of the DRL is one by Seeram [39].

The DRL has been defined by several organizations, however, only two, the ICRP and the American College of Radiology (ACR) definitions will be presented here for the sake of brevity. The ICRP [37] states that DRLs "are a form of investigation level, applied to an easily measured quantity, usually the absorbed dose in air, or tissue-equivalent material at the surface of a simple standard phantom or a representative patient." The ACR [40], on the other hand, defines a DRL as "an investigation level to identify unusually high radiation dose or exposure levels for common diagnostic medical x-ray procedures."

These formal definitions imply that the DRL is a means to reduce and/or optimize the radiation dose to the patient, and it "is to provide a benchmark for comparison, not to define a maximum or minimum exposure limit." [37]. Additionally, the ICRP states that DRLs "apply to medical exposure, not to occupational and public exposure, thus they have no link to dose limits or dose constraints… The values should be selected by professional medical bodies, and renewed at intervals that represent a compromise between the necessary stability and long-term changes in the observed dose distributions. The selected value will be specific to a country or region." [37]

DRLs are tools that can be used by radiology departments to ensure that patients are not exposed to unnecessary doses. If patients are exposed to *higher* DRL values as established by radiation protection authorities in respective countries, then corrective actions must be taken to ensure that proper radiation protection procedures are implemented in these imaging facilities can use to measure and assess radiation doses to patients for a defined set of procedures.

It is not within the scope of this Chapter to describe details of how DRLs are established however, the following points summarize existing ICRP guidelines for DRLs:

- The DRL is an advisory, not a regulatory, measure. It is not related to dose limits established for radiation workers and members of the public.
- The DRL is intended to identify high levels of radiation dose to patients.
- The DRL applies to common examinations and specific equipment.
- The DRL selection is established by professional radiation protection organizations, using a percentile point on the observed distribution for patients, and specific to a country or region. One pragmatic method is to use what has been referred to as a diagnostic reference range. While the upper level of the range is established at 75th percentile, the lower level is set at 25th percentile of the computed patient dose.
- Below a 25th percentile level, image quality may be compromised, above the 75th percentile may indicate excessive dose.

Figure 5.2 illustrates hypothetical data collected from a survey of 9 hospital CT scanner rooms showing the doses (mGy) delivered for an AP lumbar spine projection. The dose values were arranged in increasing order. The next step is to establish DRLs. In this regard the ICRP [37] recommends that "the median value (not the mean value) for the DRL quantity from each of the facilities in the survey should be used", and that national DRL values should be set as the 75th percentile (Fig. 5.2). This value becomes the reference level. It is clear from Fig. 5.2, that 2 hospitals are above this level. This means that these 2 hospitals need to engage in radiation protection activities that would bring the dose levels below the 75th percentile. For a thorough description of setting DRL values, the interested reader should refer to the ICRP Report 135 [37].

For actual DRL values for various imaging examinations, the reader should consult their respective national radiation protection organizations. For example, the ACR, the American Association of Physicists in Medicine (AAPM), and the Society for Pediatric Radiology (SPR) [40] have issued the following DRLs: for the adult PA chest (23 cm patient thickness with grid), AP abdomen (22 cm patient thickness), and the AP lumbosacral spine (22 cm patient thickness) to be 0.15, 3.4, and 4.2 mGy, respectively.

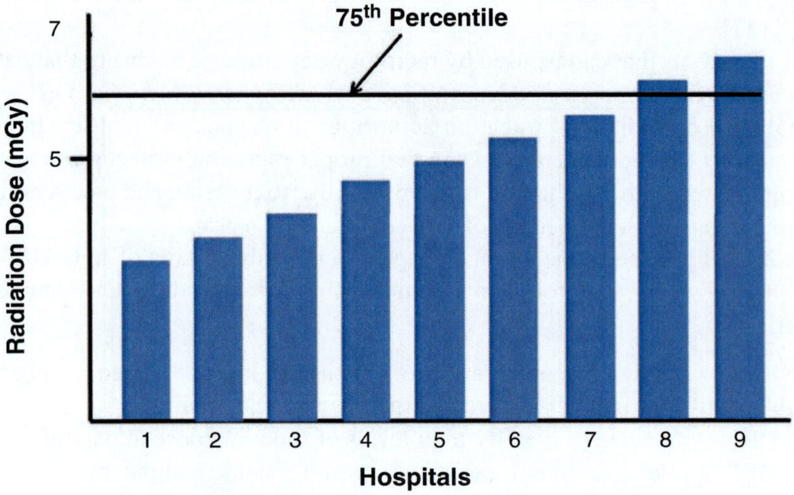

Fig. 5.2 Hypothetical data collected from a survey of 9 hospital CT scanner rooms showing the doses (mGy) delivered for an AP lumbar spine projection. The dose values were arranged in increasing order. (See text for further explanation)

5.7.2 The Use of AI-Based Image Reconstruction

The limitations of the FBP and IR image reconstruction algorithms in *Low Dose (LD) CT* imaging provides the motivation to develop algorithms that overcome these shortcomings as described in Chap. 4. In summary, for CT imaging using FBP algorithms, image quality is adequate when the dose is high; however, image noise and artifacts result when dose is lowered in an effort to reduce dose to patient and work within the ALARA philosophy. This problem is solved using iterative reconstruction (IR) algorithms, however, the literature for example has shown that in low-dose CT imaging, the noise texture produced by IR algorithms do appear blotchy, plastic-looking, or unnatural [41]. This shortcoming has provided the motivation for the development and use of AI-based image reconstruction algorithms.

In the AI domain, deep learning is a subset of machine learning, which is a subset of AI. For more details of the use of AI in medical imaging and in particular CT image reconstruction, the reader should refer to a CT textbook by Seeram [22], particularly intended for radiologic technologists. The use of deep learning image reconstruction (DLIR) has gained acceptance in CT imaging since it provides improved image quality, dose performance, and reconstruction speed compared with iterative reconstruction methods, in low-dose CT imaging.

In 2019, the US Food and Drug Administration (FDA) approved two DLIR-based, the Advanced Intelligent Clear-IQ Engine (AiCE) from Canon Medical Systems and TrueFidelity from General Electric (GE) Healthcare [31, 42]. Figure 5.3 provides an illustration of a generalized framework for the AiCE and TrueFidelity consisting of 3 steps, namely; deep learning algorithm development; training and optimization of the algorithm; and performance or verification of the algorithm.

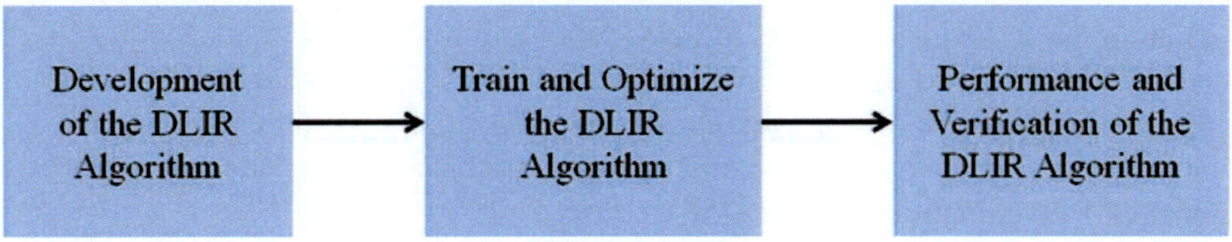

Fig. 5.3 An illustration of a generalized framework for the AiCE and TrueFidelity AI-based image reconstruction algorithms. There are essentially 3 steps, namely; deep learning algorithm development; training and optimization of the algorithm; and performance or verification of the algorithm

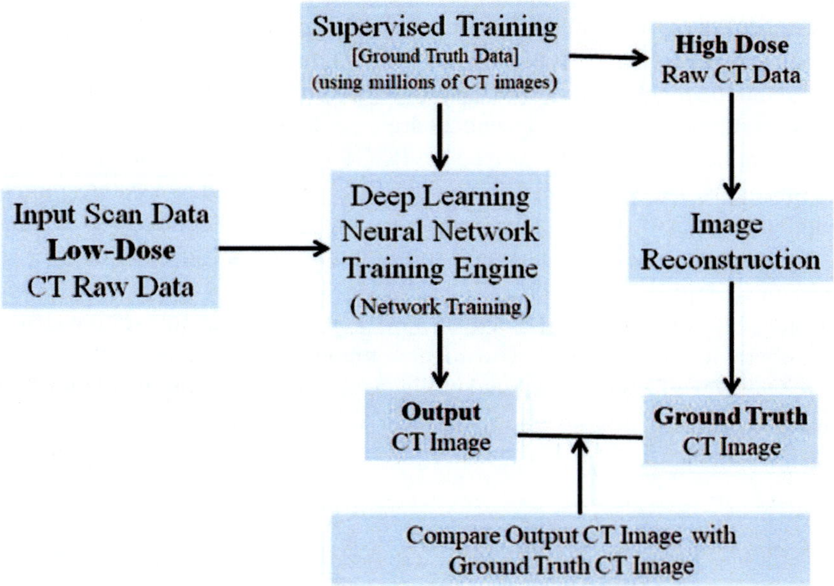

Fig. 5.4 This figure illustrates a simple overview of the deep learning process; low-dose CT data and high-dose CT data are fed into the deep learning reconstruction engine which performs the required operations to produce output images with high signal-to-noise ratio

In Fig. 5.4 illustrates a simple overview of the process; low-dose CT data and high-dose CT data are fed into the deep learning reconstruction engine which performs the required operations to produce output images with high signal-to-noise ratio.

The training process is complex, and therefore only a brief overview is given here. Duing training, output images are compared with a *reference image*, referred to as a *ground truth image* using various parameters, including image noise, noise texture, low contrast resolution, and low contrast detectability. Differences in the output image are sent to the network via backpropagation, which strengthens some equations and weakens others, and tries again. This process repeats itself until the output image is an accurate representation of the ground truth image [31].

The literature includes studies on assessing the performance of deep learning reconstruction using both test phantom and clinical patient images on noise reduction, noise texture, contrast-to-noise ratio, and low contrast detectability [31, 42–45]. These studies clearly demonstrate that deep learning reconstruction algorithms outperform iterative reconstruction algorithms on these parameters. Furthermore, they have fast reconstruction speed to meet the needs of sophisticated clinical CT examinations [31, 42–45].

5.7.3 *The Use of PCDs*

The use of PCDs as a system component of a CT scanner has been approved by the U.S. Food and Drug Administration (FDA) on September 30, 2021, for the *Naeotom Alpha* is the first commercialized Photon-Counting CT (PCCT) scanner from Siemens Healthineers. Major CT vendors are now developing PCCT Scanners.

The literature is replete with articles on PCCT scanners. For example, a Google search on August 24, 2022, for photon counting CT generated 573,000 results. It is not within the scope of this Chapter to describe the technical details of PCDs, however, the interested reader should refer to the works of Rajendran et al. [46], and Danielsson et al. [47].

The major system differences between a PCD and an Energy Integrating Detector (EID) were reviewed in Chap. 4. The major component differences are illustrated in Fig. 5.5. One type of EID is illustrated in Fig. 5.5a. This type consists of scintillation crystals (such as cadmium tungstate (CdWO4); ceramic material made of

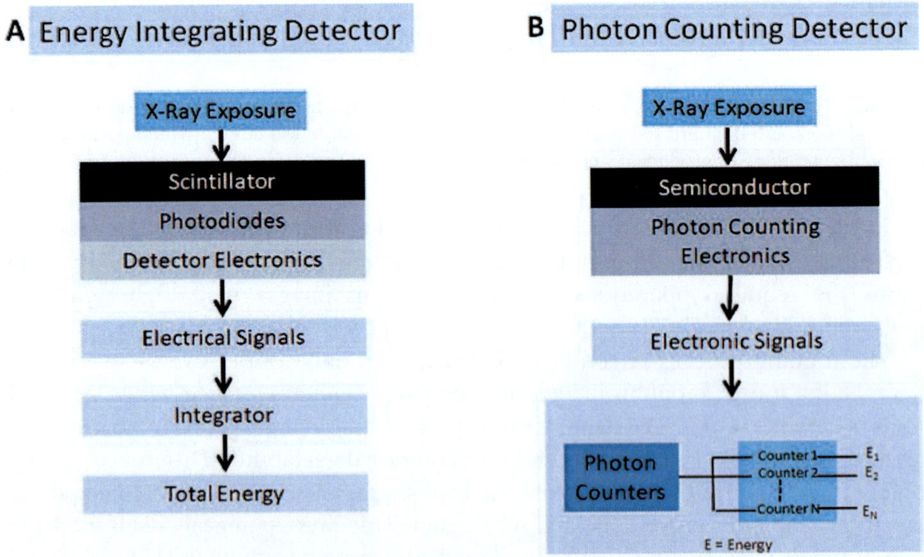

Fig. 5.5 The major system differences between a Photon Counting Detector and an Energy Integrating Detector (EID). See text for explanation

high-purity, rare-earth oxides based on doped rare-earth compounds such as yttria; and gadolinium oxysulfide ultrafast ceramic) convert x-ray photons to light photons, which then are converted to electrical signals by photodiodes. Detector electronics called application-specific integrated circuits (ASIC) digitize the signals. Digital data are subsequently fed into the CT host computer. A PCD (Fig. 5.5b) on the other hand use semiconductors such as cadmium telluride (CdTe) and cadmium zinc telluride (CZT) to convert x-ray photons directly into electron hole pairs (electric signals). These signals are fed into the photon counter electronics which provide respective energies (E_1, E_2....E_N). Digital data are then fed into the CT host computer.

In terms of dose reduction, PCCT scanners provide images at lower dose than conventional CT scanners, and provide images with higher contrast-to-noise ratio, resulting in higher resolution. For example, two studies, one by Rajendran et al. [48], and the other by Benson et al. [49] show that:

1. "The temporal bone patient images demonstrated a *dose reduction of up to 85%* compared with clinical EID-CT" [48]
2. "Temporal bone CT images obtained on a photon-counting detector CT scanner were rated as having superior spatial resolution and better critical structure visualization than those obtained on a conventional energy-integrating detector scanner, *even with a substantial dose reduction*" [49].

5.8 The Role of the Technologist in CT Dose Reduction and Optimization

There are several types of imaging professionals working in CT imaging community, each of whom play a role in responsibility in optimizing dose in CT. For example, Strauss et al. [33] and Matthews and Brennan [50] identified radiologists, physicians, medical physicists, technologists, and CT manufacturers in this regard. Strauss et al. [33] have identified 10 steps to optimize image quality and reduce radiation dose in CT. The first of these steps is to increase awareness and understanding of CT radiation dose issues among radiologic technologists.

As reviewed by Seeram [51], "the technologist is central to CT image acquisition and generally is responsible for equipment start-up procedures and patient care and communications throughout the entire examination, along with patient positioning, communications with the radiologist regarding all clinical aspects of the examination, and overall radiation protection of the patient and other staff present during the examination". As a result, the technologist plays a significant role in patient dose and image-quality optimization in CT. Seeram [51] identifies several factors that technologists must be well-versed in order to be successful in CT dose optimization strategies. These are listed in Table 5.1.

It is mandatory that in order to participate effectively in CT dose optimization through research, technologists must understand the distinction between CT dose

Table 5.1 Several factors that technologists must be well-versed in order to be successful in CT dose optimization strategies. Radiologic Technology 85 (4) 2014

1. The risks of radiation and CT dose in particular.
2. Current technical advances in CT.
3. CT dose metrics, particularly $CTDI_{vol}$, the DLP and effective dose, and associated units.
4. CT image-quality metrics such as spatial resolution, contrast resolution, noise, and artifacts.
5. The technical factors affecting patient dose in CT, including exposure technique factors (kV and mAs), Automatic Exposure Control, pitch, effective milliampere seconds, slice thickness, scan field-of-view, beam collimation, noise-reducing algorithms (iterative reconstruction algorithms), anatomical coverage, overbeaming and overranging, patient centering, and noise index
6. Scan protocols and reviewing the protocols with the radiologist on an ongoing basis with the goal of optimizing dose and image quality.
7. The prescan and postscan display of CT dose reports showing the $CTDI_{vol}$, the DLP, and effective dose.
8. Get involved with the development or implementation of a CT dose-monitoring or dose-tracking system for the CT department. Monitoring and tracking should include items such as dose capture, conversion of absorbed dose to effective dose, patient-specific storage, dose analytics, dose communication, and data export
9. Participate in research on CT patient dose and image-quality optimization. This requires a fundamental knowledge of CT equipment and dosimeter calibration, image acquisition details, observer performance measures, and appropriate statistical tools.
10. Ensure continuous professional development through relevant continuing education activities.

reduction and CT dose optimization and the application of ALARA. Dose optimization research will be described in Chap. 6.

References

1. Hricak H, Brenner DJ, Adelstein SJ, Frush DP, et al (2011). Managing radiation use in medical imaging: a multifaceted challenge. *Radiology*; 258(3):889–905. doi:https://doi.org/10.1148/radiol.10101157.
2. Bellolio MF, Heien HC, Sangaralingham LR, Jeffery MM, Campbell RL, Cabrera D, Shah ND, Hess EP (2017). Increased Computed Tomography Utilization in the Emergency Department and Its Association with Hospital Admission. West J Emerg Med;18(5):835–845. doi: https://doi.org/10.5811/westjem.2017.5.34152. Epub 2017 Jul 19. PMID: 28874935; PMCID: PMC5576619.
3. Maxwell S, Ha NT, Bulsara MK, et al (2021). Increasing use of CT requested by emergency department physicians in tertiary hospitals in Western Australia 2003–2015: an analysis of linked administrative data. BMJ;11:e043315. doi: https://doi.org/10.1136/bmjopen-2020-043315
4. Brenner DJ, Hall EJ (2007). Computed tomography: an increasing source of radiation exposure. N Engl J Med;357(22): 2277-2284.
5. Berrington de Gonzalez A, Mahesh M, Kim KP, et al (2009). Projected cancer risks from computed tomographic scans performed in the United States in 2007. Arch Intern Med;169(22):2071-2077. doi:https://doi.org/10.1001/archinternmed.2009.440.

6. Van der Molen AJ, Stoop P, Prokop M, Geleijns J (2013). A national survey on radiation dose in CT in The Netherlands. Insights Imaging;4(3):383–390. doi:https://doi.org/10.1007/s13244-013-0253-9
7. Mathews JD, Forsythe AV, Brady Z, et al (2013). Cancer risk in 680,000 people exposed to computed tomography scans in childhood or adolescence: data linkage study of 11 million Australians. BMJ; 346:1–18. doi:https://doi.org/10.1136/bmj.f2360.
8. Hendee WR, O'Connor MK (2013). Radiation risks in medical imaging: separating fact from fantasy. Radiology;264(2), 312–320. doi:https://doi.org/10.1148/radiol.12112678
9. Bushberg JT, Seibert JA, Leidholdt EM, Boone JM (2021) The Essential Physics of Medical Imaging. 3rd ed. Philadelphia, PA: Lippincott Williams & Wilkins; A Wolters Kluwer Business.
10. Sykes PJ (2020). Until There Is a Resolution of the Pro-LNT/Anti-LNT Debate, We Should Head Toward a More Sensible Graded Approach for Protection from Low-Dose Ionizing Radiation. Dose Response; 18(2):1559325820921651. doi: https://doi.org/10.1177/1559325820921651. PMI: 32425725; PMCID: PMC7218310.
11. The optimisation of radiological protection: broadening the process. ICRP Publication 101. Approved by the Commission in September 2005. Ann ICRP. 2006;36(3):69–87.
12. Miglioretti DL, Johnson E, Williams A, et al (2013). The use of computed tomography in pediatrics and the associated radiation exposure and estimated cancer risk. JAMA Pediatr; 167(8):700–707. doi:https://doi.org/10.1001/jamapediatrics.2013.311
13. Pearce MS, Salotti JA, Little MP, et al (2012). Radiation exposure from CT scans in childhood and subsequent risk of leukemia and brain tumors: a retrospective cohort study. Lancet; 80(9840):499–505. doi:https://doi.org/10.1016/S0140-6736(12)60815-0
14. Hall EJ, Brenner DL (2014) Cancer risks from diagnostic radiology. British Journal of Radiology Vol. 81, No. 965
15. Shao YH, Tsai K, Kim S, Wu YJ, Demissie K. (2020). Exposure to Tomographic Scans and Cancer Risks. JNCI Cancer Spectr;4(1):pkz072. doi: https://doi.org/10.1093/jncics/pkz072. PMID: 32337490; PMCID: PMC7050152.
16. Moghadama N, Rehani M M, Nassiri MA (2021) Assessment of patients' cumulative doses in one year and collective dose to population through CT examinations. European Journal of Radiology, Volume 142, September 2021, 109871. doi:https://doi.org/10.1016/j.ejrad.2021.109871
17. Zewdea N, Riab F, Rehani MM (2022) Organ doses and cancer risk assessment in patients exposed to high doses from recurrent CT exams. European J of Radiol; 149: 110224.
18. Hong J-Y, Han K, Jung J-H Jung, Kim S (2019). Association of Exposure to Diagnostic Low-Dose Ionizing Radiation with Risk of Cancer Among Youths in South Korea. JAMA Network Open;2(9):e1910584. doi:https://doi.org/10.1001/jamanetworkopen.2019.10584
19. Sakane H, Ishida M, Shi L, et al (2020). Biological Effects of Low-Dose Chest CT on Chromosomal DNA. Radiology;295(2):439-445. DOI: https://doi.org/10.1148/radiol.2020190389. PMID: 32154776.
20. Radiological protection and safety in medicine (2007). Recommendations of the International Commission on Radiological Protection. Publication No. 103. Ann ICRP. 2007; 37:2–4.
21. McNitt-Gray MF (2002). AAPM/RSNA Physics Tutorial for Residents: Topics in CT: Radiation dose in CT. Radiographics;22(6):1541–1553.
22. Seeram E (2023). Computed Tomography: Physical Principles, Patient Care, Clinical Applications, and Quality Control. St Louis, MO. Elsevier
23. Wolbarst AB, Capasso P, Wyant AR (2013). Medical Imaging: Essentials for Physicians. Hoboken, NJ: Wiley-Blackwell.
24. Goo HW (2012). CT radiation dose optimization and estimation: an update for radiologists. Korean J Radiol;13(1):1–11. doi:https://doi.org/10.3348/kjr.2012.13.1.1.
25. Reid J, Gamberoni J, Dong F, Davros W (2010). Optimization of kVp and mAs for pediatric low dose simulated abdominal CT: Is it best to base parameter selection on object circumference? AJR Am J Roentgenol;195(4):1015–1020. doi:https://doi.org/10.2214/AJR.09.3862.

26. Maldjian PD, Goldman AR (2013). Reducing radiation dose in body CT: primer on dose metrics and key CT technical parameters. AJR Am J Roentgenol; 200(4):741–747. doi:https://doi.org/10.2214/AJR.12.9768
27. Christner JA, Zavaletta VA, Eusemann CD, Walz-Flannigan AL, McCollough CH (2010). Dose reduction in helical CT: dynamically adjustable z-axis x-ray beam collimation. AJR Am J Roentgenol;194(1):W49–W55. doi:https://doi.org/10.2214/AJR.09.2878.
28. Toth T, Ge Z, Daly MP (2007). The influence of patient centering on CT dose and image noise. Med Phys;34(7):3091–3101.
29. Beister M, Kolditz D, Kalender WA (2012). Iterative reconstruction methods in x ray CT. Phys Med;28(2):94–108. doi:https://doi.org/10.1016/j.ejmp.2012.01.003.
30. Kaza RK, Platt JF, Goodsitt MM, et al (2014). Emerging techniques for dose optimization in abdominal CT. Radiographics; 34(1):4–17. doi:https://doi.org/10.1148/rg.341135038.
31. Hsieh J, Nett B, Yu Z, Sauer K, Thibault J-B, Bouman CA (2013). Recent advances in CT image reconstruction [published online January 16, 2013]. Curr Radiol Rep. 2013. doi:https://doi.org/10.1007/s40134-012-0003-7.
32. "Reduction", "Optimization". Merriam-Webster Dictionary. Online English Language Dictionary. www.merriam – webster.com/dictionary. Accessed August 2022.
33. Strauss KJ, Goske MJ, Kaste SC, et al (2010). Image gently: Ten steps you can take to optimize image quality and lower CT dose for pediatric patients. AJR Am J Roentgenol;194(4):868–873. doi:https://doi.org/10.2214/AJR.09.4091
34. Karami V, Gholami M (2018) Addressing as low as reasonably achievable (ALARA) in pediatric computed tomography (CT) procedures, J Res Med Dent Sci; 6 (5):104–114
35. Sohaib SA, Peppercorn PD, Horrocks JA, Keene MH, Kenyon GS, Reznek RH (2001). The effect of decreasing mAs on image quality and patient dose in sinus CT. Br J Radiol;74(878):157–161.
36. Russell MT, Fink JR, Rebeles F, Kanal K, Ramos M, Anzal Y (2008). Balancing radiation dose and image quality: clinical applications of neck volume CT. Am J Neuoradiol;29(4):727- 731. doi:https://doi.org/10.3174/ajnr.A0891.
37. ICRP Publication 135 (2017). Diagnostic Reference Levels in Medical Imaging. Annals of the ICRP Volume 46, Issue 1, pages 1–144
38. Seeram, E. and Brennan, P (2006). Diagnostic Reference Levels in Radiology. Radiologic Technology; Vol. 77/No. 5; 373–384
39. Seeram, E. (2020). Rad Tech's Guide to Radiation Protection, 2e. Oxford: Wiley.
40. American College of Radiology ACR-AAPM-SPR (2020). Practice parameter for diagnostic reference levels and achievable doses in medical x-ray imaging, 2018. https://www.acr.org/–/media/ACR/Files/Practice-Parameters/Diag-Ref-Levels.pdf (accessed 30 August 2020)
41. Geyer LL, Schoepf UJ, Meinel FG, et al (2015). State of the art: iterative CT reconstruction techniques. *Radiol*;276(2):339–357. doi:https://doi.org/10.1148/radiol.2015132766
42. Boedeker K (2019). AiCE deep learning reconstruction: bringing the power of ultra-high resolution CT to routine imaging. *Canon Med Syst*; 2:28–33.
43. Shan H, Zhang Y, Yang Q, et al. 3D convolutional encoder–decoder network for low-dose CT via transfer learning from a 2D trained network. *IEEE Trans Med Imaging*. 2018;37(6):1522–1534. doi:https://doi.org/10.1109/TMI.2018.2832217
44. Shan H, Padole A, Homavounieh F, et al. Competitive performance of a modularized deep neural network compared to commercial algorithms for low-dose CT image reconstruction. *Nat Mach Intell*. 2019;1:269–276. doi:https://doi.org/10.1038/s42256-019-0057-9
45. Akagi M, Nakamura Y, Higaki T, et al. Correction to: deep learning reconstruction improves image quality of abdominal ultra-high-resolution CT. *Eur Radiol*. 2019;29(8):4526–4527. doi:https://doi.org/10.1007/s00330-019-06249-x
46. Rajendran K, Petersilka M, Henning A, Shanblatt ER, Schmidt B., Flohr TG, Ferrero A, Baffour F, Diehn FE, Yu L, Rajiah P, Fletcher JG, Leng S., McCollough CH (2021). First Clinical Photon-counting Detector CT System: Technical Evaluation. Radiology, Published Online, doi:https://doi.org/10.1148/radiol.2125794

47. Danielsson M, Persson M, and Sjölin M (2021). Photon-counting x-ray detectors for CT. Phys. Med. Biol. 66; 03TR01, doi:https://doi.org/10.1088/1361-6560/abc5a5
48. Rajendran K, Voss BA, Zhou W, Tao S, DeLone DR, Lane JI, Weaver JM, Carlson ML, Fletcher JG, McCollough CH, Leng S. Dose Reduction for Sinus and Temporal Bone Imaging Using Photon-Counting Detector CT With an Additional Tin Filter. Invest Radiol. 2020 Feb;55(2):91–100. doi: https://doi.org/10.1097/RLI.0000000000000614. PMID: 31770297; PMCID: PMC8522262.
49. Benson JC, Rajendran K, Lane JI, Diehn FE, Weber NM, Thorne JE, Larson NB, Fletcher JG, McCollough CH, Leng S. A New Frontier in Temporal Bone Imaging: Photon-Counting Detector CT Demonstrates Superior Visualization of Critical Anatomic Structures at Reduced Radiation Dose. AJNR Am J Neuroradiol. 2022 Apr;43(4):579–584. doi: https://doi.org/10.3174/ajnr.A7452. Epub 2022 Mar 24. PMID: 35332019; PMCID: PMC8993187.
50. Matthews K, Brennan P. Optimization of x-ray examinations: general principles and an Irish perspective. *Radiography*. 2009; 15:262–268.
51. Seeram E (2014). Computed Tomography Dose Optimization. Radiologic Technology, 85 (4), p 655–671

Chapter 6
Optimization Research in Medical Imaging

Keywords Research · Scientific method · Dose/image quality optimization · Image quality descriptors · Image quality assessment tools · Observer · Observer performance methods · Visual grading analysis · Receiver operating characteristics

6.1 Introduction

Research is based on science (a systematic and organized body of knowledge in any area of inquiry) and is conducted using the "*scientific method*" [1]. There are at least four primary tasks in research, based on the scientific method, namely; *identification of a problem*, *data collection to help answer the question posed by the problem*, *data analysis*, and last, *provide a solution to the problem*. These four tasks require a thorough and defined set of steps that ensure a valid solution to the problem. The major steps are usually outlined as follows:

- Problem identification
- Literature review
- Research purpose statement
- Research Objectives, Questions and Hypothesis
- Research methodology and research design
- Data Collection
- Data analysis and interpretation
- Research report and evaluation

Research in medical imaging, and in particular, on dose/image quality optimization is complex and it is based on the scientific method [2]. A dose/image quality study requires that images be obtained at doses ranging from low to high. Subsequently, observers assess these images to determine the optimum dose, that is the lowest dose without compromising the image quality. In particular, McCollough et al. [3], point out that image quality assessment is a complex process and requires significant attention to quantitative and qualitative measures. Furthermore, McCollough

E. Seeram, *Dose Optimization in Digital Radiography and Computed Tomography*,
https://doi.org/10.1007/978-3-031-22871-1_6

et al. [3] stress that quantitative metrics such as image noise and observer performance, be considered.

6.1.1 Strategies Used in Dose/Image Quality Optimization Research

The literature includes formal approaches used to guide dose/image quality optimization research [4]. One important and significant seminal work is a special issue of the scientific journal *Radiation Protection Dosimetry* [5]. This issue is dedicated to the nature and strategies for optimization research in medical imaging, including radiography, fluoroscopy, mammography, and computed tomography (CT).

Several studies featured in this special issue have identified at least four essential requirements that must be met for an effective and thorough study on dose/image quality optimization research study [5] These requirements include:

1. Patient safety must be a primary consideration
2. The level of image quality for the diagnostic task must be determined
3. Use dose levels ranging from high to low keeping in mind that image quality must not be compromised.
4. Valid research methodologies [6, 7] must be used for:
    - Dosimetry measurements
    - Image acquisition
    - Evaluation of image quality obtained at all dose levels, by human observers keeping in mind the
    - Nature of the detection task

Additional Considerations

There are several additional requirements as described in a paper by Seeram [8] published in Radiologic Technology, for dose/image quality optimization research in CT, that are considered minimum essential elements. These include:

1. Calibration of the CT imaging system, to ensure that the system performance is reliable and consistent during dose measurements
2. Calibration of the dosimeters used to measure doses during imaging acquisition
3. Appropriate anthropomorphic phantoms and image quality test tools must be used for image acquisition
4. Reliable and approved methodology for dose measurements
5. Assessment of images should occur at least in two phases:

    A. Expert human observers assess image quality on a defined and specific criterion, for example, the appearance of image noise

B. The assessment of images recorded with various levels of dose should be done using the same set of observers using established, valid, and reliable observer performance methods
C. The selection of the appropriate observer performance method may be based on the nature of the detection task [9]
D. Visual Grading Analysis (VGA) is a useful tool if the detection task is based on the visualization of anatomic structures. VGA is based on the assumption that an observer's ability to see and evaluate normal anatomy correlates to the ability to evaluate pathology, or abnormal findings [9]
E. Receiver Operating Characteristic (ROC) method is a useful tool for the assessment of pathology detection [9]

A brief overview of VGA and ROC will be provided later in the Chapter.

6.1.2 The Nature of Image Quality

Dose optimization research seeks to select the lowest possible dose to the patient without compromising the image quality required to make a diagnosis by the human observer. In this regard then, image quality characteristics are important and must be considered, as an integral component in any research study.

Image Quality Descriptors

In a recent paper on "The assessment of image quality and diagnostic value in X-ray images: a survey on radiographers' reasons for rejecting image", Kjelle and Chilanga [10] stress the following: "The quality of a radiographic image influences diagnostic accuracy and subsequent clinical management of the patient [11]. A radiographic image is accepted as good quality when certain technical qualities are satisfied, and the image considered of diagnostic value" [12]. Good image quality in diagnostic imaging can be assessed mainly by at least three image quality factors, namely; Image quality in digital imaging includes *spatial resolution*, *contrast-to-noise resolution,* and *image noise*. These will be reviewed below.

Dose optimization not only address reduction of radiation dose according to the ALARA principle, it focusses also on image quality. In CT imaging for example, there are at least four essential image quality parameters that relate to radiation dose [13]. These include spatial resolution, contrast-to-noise resolution, image noise, and they have been described in detail in the literature by several authors [3, 4, 6–9]. The first three have been outlined in Chap. 5. *In review,* Goo [13] provides a useful summary as follows:

- *Image Noise* is a "random variation of CT numbers" [13] and it is "inversely proportional to the square root of radiation dose; inversely proportional to tube

voltage; inversely proportional to the fourth power of the spatial resolution; influenced by section thickness and reconstruction algorithm" [13]
- *Contrast-to-noise resolution* is the "ability to distinguish between different CT numbers" [13]. The relationship with dose is "not significantly changed at different tube voltages in most materials except for some with high atomic numbers such as iodine (increased at low tube voltages), low contrast resolution greatly influenced by image noise level" [13].
- *Spatial resolution* is the "ability to distinguish small details of objects" [13]. Its relationship with dose is that it is "inversely proportional to focal spot size, detector collimation; and influenced by the reconstruction algorithm" [13].

Furthermore, Goo [13] points out that "the required image quality of CT differs somewhat among different diagnostic tasks. Consequently, the required CT radiation dose is also fairly diverse and thus should be tailored according to clinical indications. For example, very low dose CT can be used for the identification of high-contrast lesions, such as urinary stones, colonic polyps (virtual colonography), or lung nodules."

6.1.3 Image Quality Assessment Tools for Clinical CT Images: An Overview

Chipiga et al. [14], Zarb et al. [15] and Seeram et al. [7], have described image quality assessment tools for optimization of CT images. In particular, two approaches have been identified;

1. Objective physical measures
2. Observer and diagnostic performance tests

While the former is based on the evaluation of physical and technical parameters of the image, the latter is based on assessment of clinical images of patients, and/or anthropomorphic phantoms.

Objective Physical Measures

The major physical characteristics of a digital imaging system (radiographic or CT) are those that relate to the performance of the imaging system. These characteristics and performance measures include the modulation transfer function, the Signal-to-Noise Ratio (SNR), the Wiener spectrum (also known as the noise power spectrum), and the Detective Quantum Efficiency (DQE). While these are useful and significant tools to evaluate image quality, they do not relate to all components of the imaging chain [16–18]. Furthermore, Tapiovaara [19], in a review of the relationships between physical measurements and human observer evaluation of image quality, emphasized that “exact physical measurements cannot supersede subjective evaluation in judging the acceptability of clinical images.”

Clinical medical imaging involves subjective evaluation of patient images using human observers, usually radiologists. The technologist however, is the first observer to assess image quality before images are sent to radiologists for diagnostic interpretation. In general, while the major task of the technologist is to evaluate the visualization and reproduction of relevant anatomical structures, the major task of the radiologist is detection of subtle lesions (detection of pathology). The procedures used for these two purposes fall into the category of observer performance methods.

Observer Performance Methods

Mansson [16], in describing various approaches to evaluating clinical image quality, identifies 4 tasks that the observer must consider. These include:

1. The differences between varying states of diseases and health should be discerned (perceived or recognized)
2. Relevant anatomical structures and features should be described accurately
3. Abnormalities should be classified accurately, and
4. Relevant anatomy on images should be distinguished reliably

While the first three goals fall within the viewing and interpretation task of the radiologist, the last one (visualization and inclusion of relevant anatomical structures) belongs to the viewing task of the radiologic technologist. To measure these tasks, requires the use of *observer performance methods.*

There are at least two classes of observer performance methods used to asses clinical diagnostic image quality, not only for the visualization and reproduction of normal anatomical structures but also for lesion detection. These include:

1. Visual Grading Analysis (VGA) used in the visualization and reproduction of normal anatomical structures
2. Receiver operating characteristic (ROC) analysis is used in lesion detection tasks.

The major features of each of these will be described in the next sections of this Chapter.

6.1.4 Receiver Operating Characteristics: A Brief Overview

The ROC analysis is a commonly used approach to measure the performance of the human observer on the task of the detection of lesions in patient images in medical imaging [20, 21]. ROC analysis is complex and requires a firm understanding of several integral elements on which the analysis is based. For example, in order to identify lesions (abnormalities) in an image, calls for an understanding of two terms, namely; sensitivity and specificity. While *sensitivity* refers to the detection of abnormalities when present, *specificity* refers to ruling out abnormalities that are absent (false positive rate = FPR or 1). These two factors are fundamental to ROC analysis.

If the sensitivity (on the y-axis) is plotted as a function of specificity (on the x-axis), an ROC curve is obtained, which shows the relationship between sensitivity and FPR. Zarb et al. [15], summarises the plot as follows: "The area under the ROC curve is the most common measure of the accuracy of the test. A test with an ROC curve area of 1 is perfectly accurate with a sensitivity of 1 and FPR of 0. On the other hand, an ROC curve area of 0 is perfectly inaccurate where all patients are incorrectly given positive results. In practice the lower limit for the RO C curve area is given at 0.5 which is equivalent to guessing the patient's result hence a 50% chance of getting the correct interpretation. ROC curve areas above 0.5 have some ability to discriminate between patients having a disease and those who do not".

It is not within the scope of this Chapter to outline the details of ROC analysis, and therefore the reader is encouraged to refer to McEntee [22], for a comprehensive and further updates on ROC analysis, including its variants, such as for example; the free-response ROC, localization ROC, and alternative free-response ROC, and so forth. In terms of the *shortcomings of ROC analysis*, Zarb et al. [15], states the following: "The data has to be divided in only 2 categories, normal or abnormal; a large number of images with subtle pathology are required to produce statistically significant results; restricted to one observer per report per image; ROC method does not cater for multiple abnormalities within the same image; abnormality localization is not considered and hence an image may be marked as abnormal for the wrong reason". Furthermore, an important consideration, however, and within the context of this book, "concerns have been raised regarding their clinical relevance" [16, 17]. Additionally, Bath [17] notes that "conducting these types of studies may be cumbersome, because they are based on the establishment of truth for all cases and normally require a large number of cases in order to produce statistically significant results, which means that the reliability is relatively low. For these reasons, ROC-related methods may not be the method of choice for the local optimization task at a radiation department where the intention is to find the optimal image processing setting or (dose level) for a given examination".

6.1.5 Visual Grading Analysis: An Overview

A number of authors [4, 6, 9, 15–18, 23] have demonstrated that visual grading using VGA is also another valid useful tool for evaluating image quality. VGA is well-established where the observer's overall goal is to subjectively assess image quality, based on the visualization and reproduction of anatomical structures at different dose levels. Sund et al. [18] point out that a critical assumption of VGA is that "the visibility of normal anatomy correlates to the detectability of pathological structures." Furthermore, Bath [17] identifies several reasons for using visual grading in image quality assessment studies. These include:

1. It is assumed that the validity of VGA studies is high for assessing visualization of relevant anatomical structures.
2. "Visual grading methods have in special cases been shown to agree both with detection studies using human observers and with advanced calculations of the physical image quality. This is important, and validates in some way the assumption that the possibility to detect pathology correlates to the reproduction of anatomy—the basic ideas of visual grading. Discrepancies between the methods have been reported but have been explained with the different tasks for the methods rather than low validity for visual grading" [17].
3. VGA research is simple to carry out (compare to ROC studies). "How to perform visual grading studies has been extensively described and the learning threshold for conducting such studies is low" [17]
4. VGA studies do not demand too much time on the part of the observer.

The use of VGA however, has been subject to concerns that since subjectivity is one of the tasks of the observer in assessing images, it could lead to bias. To address with this bias problem, a set of formal criteria have been developed by a group of expert radiologists and medical physicists. These criteria are the *European Guidelines on Quality Criteria for Diagnostic Radiographic Images* and were published by the Commission of European Communities in 1996 [24].

It is not within the scope of this Chapter to present details of such criteria therefore, the reader is encouraged to refer to reference 24 for full details.

VGA Methods

The literature identifies 2 common VGA approaches, and as noted earlier, VGA requires that the observer assess image quality in a subjective manner using an *absolute* or a *relative rating* scale. While the former requires the observer to rate his or her opinion about the visibility of the anatomical structures, the latter requires the observer to rate the visibility of each defined anatomical structure. For example, with the absolute rating scale the structure in the image is: not visible; poorly reproduced; adequately reproduced; or very well reproduced. With the relative rating scale, the reproduction of the structure in the image is ___; the reproduction of the corresponding structure in the reference image. The blank space would be filled in with one of the selected choices as follows: much worse than (−2); worse than (−1); the same as (0); better than (+1); or much better than (+2).

An important consideration in conducting a VGA study is that in using the relative approach, the reference image should always be presented side by side on the same monitor used to display the test images to ensure that these images are displayed with the same monitor brightness and contrast [25, 26].

6.1.6 Example of a Dose Optimization Study in Digital Radiography

A study that illustrates all the *elements of a typical dose optimization study* in medical imaging, based on the scientific method is one by Seeram et al. [27]. The study used a *descriptive correlational research design* to investigate "*the optimization of a of the exposure indicator (EI) of a computed radiography imaging system as a radiation dose management strategy*". A brief summary by Seeram et al. [2] *is* as follows:

1. "The study compared the optimized EI with the manufacturer's recommended values for the anteroposterior (AP) pelvis and AP lumbar spine projections. One of the study's objectives was to correlate the milliampere seconds (mAs) and associated EIs with the entrance skin dose using a range of mAs settings while holding the kilovoltage peak (kVp) constant.
2. A noninterventional research design was used to relate or associate variables in a predictable manner as opposed to an interventional approach, which is typical in experiments. The noninterventional approach used a correlational research method to measure the degree of association (or relation) between 2 or more variables using statistical procedures. Entrance skin dose and EI were the dependent variables, and mAs was the independent variable.
3. The measurement of the entrance skin dose for the AP pelvis and the AP lumbar spine using exposure technique factors including mAs and kVp that the manufacturer suggested. Researchers obtained the entrance skin dose measurements with a calibrated dosimeter (Unfors Instruments) free-in-air (ie, an air environment free of any absorbing or scattering objects) for the AP pelvis and the AP lumbar spine. For all 9 of the mAs settings, researchers calculated the mean milligray (mGy) per mAs setting for the AP pelvis and AP lumbar spine.
4. Descriptive statistics (e.g., such as sample size, mean, standard deviation, and range) were computed for the dosimetry and EI data. In addition, the Pearson correlation was applied to examine the relationship between the entrance skin dose and the EI as well as the dose and mAs. Data was graphed and plotted the mean dose for the AP pelvis and the AP lumbar spine as a function of mAs. Furthermore, the inverse EI, the mAs, and the dose were compared. The inverse EI shows a strong positive linear relationship ($r = 0.999$) for both the AP pelvis and the AP lumbar spine.
5. At doses of 5 mGy, 10 mGy, and 20 mGy, the EIs were 400, 200, and 100, respectively. When researchers plotted the inverse EI as a function of mAs and plotted the dose as a function of the inverse EI, they saw a strong positive linear relationship ($r = 0.999$) in both cases. Instead of displaying the S number on an image (e.g., 100), the inverse S number (e.g., 0.01) should be displayed to better aid technologists in understanding the relationship between the dose and the patient (as opposed to the image plate) as well as the S number.

6. The results of this study show that technologists can optimize the dose and image quality in computed radiography imaging using the mAs exposure technique factor and associated EIs. Specifically, the dosimetry phase of this investigation showed a strong positive linear relationship ($r = 0.999$) between mAs and dose, mAs and the inverse EI, and the inverse EI and dose for the AP pelvis and AP lumbar spine. Under the controlled conditions researchers used in this study for dose optimization, the EI values were stable. The manufacturer's reference values of 25 mAs (EI = 86) for the AP pelvis and 50 mAs (EI = 88) for the AP lumbar spine were optimized to 16 mAs for the AP pelvis (EI = 136) and 32 mAs for the AP lumbar spine (EI = 139). Seven expert observers, who have at least 10 years of experience teaching radiographic technique and positioning in classrooms, laboratories, and hospitals, performed an image quality assessment and determined that the manufacturer's recommended dose can be reduced by 36% for the AP pelvis and AP lumbar spine without compromising image quality."

In conclusion, dose optimization in digital radiography and computed tomography is an important responsibility of all those who play an active role in the imaging of patients. It simply means that doses should be kept as low as is reasonably achievable so as not to compromise the image quality needed for diagnosis. This is the ALARA principle of the ICRP and other international and national radiation protection organizations.

References

1. Bhattacherjee, A. (2012). Social science research: Principles, methods, and practices. In Textbooks collection. 3. http://scholarcommons.usf.edu/oa_textbooks/3. University of South Florida
2. Seeram E, Davidson R, England A, McEntee M (Eds) (2022) Research in Medical Imaging and Radiation Sciences. Switzerland AG, Springer Nature.
3. McCollough C, Bruesewitz MR, Kofler JM Jr. (2006). CT dose reduction and dose management tools: overview of available options. Radiographics; 26(2):503–512.
4. Mattsson S, ed (2005) Optimization strategies in medical x-ray imaging. Radiat Protect Dosimetry; 114(1–3):1–3. doi:https://doi.org/10.1093/rpd/nch580.
5. Radiation Protection Dosimetry. Optimization Strategies in Medical X-Ray Imaging, Vol 114, No 1–3; 2005
6. Zarb F, Rainford L, McEntee MF (2010). Image quality assessment tools for optimization of CT images. Radiography; 16:147–153.
7. Seeram E, Davidson R, Bushong S, Swan H (2014). Image Quality Assessment Tools for Radiation Dose Optimization in Digital Radiography: An Overview. Radiologic Technology; Volume 85, Number 5, 555–562
8. Seeram E (2014) Computed Tomography Dose Optimization. Radiologic Technology; Vol 85, No 6, 655–671
9. Tingberg A, Båth M, Håkansson M, et al (2005). Evaluation of image quality of lumbar spine images: a comparison between FFE and VGA. *Radiat Protect Dosimetry*; 114(1–3):53–61.
10. Kjelle, E., Chilanga, C. The assessment of image quality and diagnostic value in X-ray images: a survey on radiographers' reasons for rejecting images. *Insights Imaging* **13**, 36 (2022). https://doi.org/10.1186/s13244-022-01169-9

11. Carmichael JHE, Maccia C, Moores BM et al (2000) European guidelines on quality criteria for diagnostic radiographic images. EU Publication
12. Maccia C, Moores BM, Wall BF (1997) The 1991 CEC trial on quality criteria for diagnostic radiographic images: detailed results and findings. EN Luxembourg Office for Official Publication of the European Communities
13. Goo HW (2012). CT radiation dose optimization and estimation: an update for radiologists. Korean J Radiol;13(1):1–11.
14. Chipiga LA, Berkovich GV, Vodovatov AV, Trufanov GE (2021). Comparison of different approaches for image quality assessment in computer tomography. American Institute of Physics (AIP) Conference Proceedings 2356, 020005 (2021); https://doi.org/10.1063/5.0052960
15. Båth M, Månsson LG (2007). Visual grading characteristic (VGC) analysis: a non-parametric rank invariant statistical method for image quality evaluation. *Br J Radiol*;80(951):169–176.
16. Mansson L (2005). Methods for the evaluation of image quality: a review. Radiat Prot Dosimetry; 90(1e2):89e99
17. Båth M (2010) Evaluating imaging systems: practical applications. *Radiat Prot Dosimetry*; 139(1–3):26–36. doi:https://doi.org/10.1093/rpd/ncq007.
18. Sund P, Båth M, Kheddache S, Månsson LG (2004). Comparison of visual grading analysis and determination of detective quantum efficiency for evaluating system performance in digital chest radiography. *Euro Radiol*;14(1):48–58.
19. Tapiovaara MJ (2008). Review of relationships between physical measurements and user evaluation of image quality. *Radiat Prot Dosimetry*; 129(1–3):244–248. doi:https://doi.org/10.1093/rpd/ncn009.
20. Metz C (1986). ROC methodology in radiologic imaging. Invest Radiol; 21:720e33.
21. Chakraborty D (2006) ROC curves predicted by a model of visual research. Phys Med Biol; 51:3463e82.
22. McEntee M (Chapter 6, In Seeram E, Davidson R, England A, and McEntee M (eds) (2022) Research in Medical Imaging and Radiation Sciences. Springer
23. Precht H, Hansson J, Outzen C, Hogg P, Tingberg A (2019) Radiographers' perspectives on Visual Grading Analysis as a scientific method to evaluate image quality., Radiography; Volume 25, Supplement 1, Pages S14–S18
24. Commission of the European Communities (1996). European guidelines on quality criteria for diagnostic radiographic images. European Commission; Brussels, Belgium.
25. Tingberg AM (2000). Quantifying the Quality of Medical X-ray Images: An Evaluation Based on Normal Anatomy for Lumbar Spine and Chest Radiography [dissertation]. Lund, Sweden: Lund University, Department of Radiation Physics.
26. Gorham S, Brennan PC (2010). Impact of focal spot size on radiologic image quality: a visual grading analysis. Radiography;16(4):304–313. doi:https://doi.org/10.1016/j.radi.2010.02.007
27. Seeram E, Davidson R, Bushong S, Swan H. Optimizing the exposure indicator as a dose management strategy in computed radiography. Radiol Technol. 2016;87(4):380–391.

Chapter 7
Review Questions

Answer the following questions to check your understanding of the materials studied.

Chapter 1

1. In which of the following effects, a threshold dose exists?

    A. Leukemia
    B. Cancer
    C. Cataracts
    D. Hereditary effects

2. There is no threshold dose for:

    A. Leukemia.
    B. Cataracts
    C. Skin burns
    D. Tissue damage

3. Which of the following data sources show that there is statistically significant evidence of increased cancers in Japanese survivors who received doses of 100 mSv and higher?

    A. RERF data
    B. ICRP data
    C. NCRP data
    D. LNT hypothesis

4. The ICRP suggests that justification addresses:

    A. The concept of a good image
    B. Low dose to the patient
    C. Image quality optimization
    D. The concept of net benefit

E. Seeram, *Dose Optimization in Digital Radiography and Computed Tomography*,
https://doi.org/10.1007/978-3-031-22871-1_7

5. Which of the following radiation protection guideline relates to the ALARA philosophy?
   A. Optimization
   B. Dose limitation
   C. Justification
   D. Shielding
6. Optimization takes into consideration all of the following except:
   A. Imaging equipment and its adequacy
   B. Technical parameters of the examination,
   C. Dose limits for occupationally exposed individuals
   D. Diagnostic reference levels
7. The phenomenon of an increase in the exposure over time as technologists use these DR imaging systems is referred to as:
   A. Dose creep or exposure creep
   B. Dose limits
   C. Deterministic effects
   D. Stochastic effects
8. Which of the following imaging modality contributes to the highest collective amount of radiation exposure in the United States, as of 2011?
   A. Digital radiography
   B. Computed tomography
   C. Digital fluoroscopy
   D. Digital mammography
9. The term used to refer to "an act, process, or methodology of marking something (as a design, system, or decision) as fully perfect, functional, or effective as possible" is:
   A. Reduction
   B. ALARA philosophy
   C. Radiation protection
   D. Optimization
10. The exposure indicator (EI) and the deviation index (DI) are technical factors used in:
    A. Digital Fluoroscopy
    B. Computed Radiography
    C. Flat Panel Digital Radiography
    D. B and C are correct

Chapter 2

11. A linear response of a digital detector to radiation exposure means that:

    A. Digital image processing ensures that the density and contrast of the displayed image appear acceptable to the observer, whether the exposure is low or high
    B. A specific exposure is needed to produce an acceptable image
    C. The digital detector has better spatial resolution that film-screen (F-S) detectors
    D. The contrast resolution of the digital detector is less that F-S detectors

12. The exposure indicator (EI) in digital radiography refers to:

    A. An image with acceptable contrast and density
    B. A number to identify the use of the correct detector
    C. A number to indicate that the correct exposure technique factors have been used
    D. A repeat exposure in required

13. The current visual cue on a digital image that shows the use of the correct exposure technique factors is the:

    A. EI
    B. Deviation Index (DI)
    C. Sharpness of the image
    D. Correct contrast of the image structures

14. A digital radiography imaging system based on the principle of photostimulable luminescence is:

    A. Flat-panel digital radiography
    B. Image-intensifier based digital fluoroscopy
    C. Computed radiography imaging system
    D. All are correct

15. Which photostimulable phosphors are typically used in CR imaging plates?

    A. Cesium bromide
    B. Barium fluorobromide
    C. Barium Fluoroiodide
    D. All are correct

16. An indirect conversion flat-panel digital radiography (FPDR) detector uses:

    A. A photocathode to convert x-rays to light
    B. An analog-to-digital converter to convert light to electrical signal
    C. Selenium to convert x-rays to electrical signal
    D. A scintillator to convert x-rays to light

17. A direct conversion FPDR detector uses:
    A. A photocathode to convert x-rays to light
    B. An analog-to-digital converter to convert light to electrical signal
    C. Selenium to convert x-rays to electrical signal
    D. A scintillator to convert x-rays to light
18. The preprocessing feature characteristic of digital radiography imaging systems is mainly used to:
    A. Identify, correct, and scale the raw image data prior to image display for the purpose of viewing and interpretation
    B. Enhance image contrast
    C. Enhance density and sharpness of the raw data
    D. Allow the technologist to use windowing to change image contrast and brightness
19. Which of the following digital postprocessing operation changes the sharpness of the displayed image?
    A. Window width processing
    B. Window level processing
    C. Spatial frequency processing
    D. All are correct
20. Which of the following refers to the expected value of the exposure index when exposing the x-ray image receptor properly?
    A. EI
    B. Target EI (EI_T)
    C. DI
    D. Volume of interest (VOI)
21. The number quantifying the deviation of the actual exposure index from a target exposure index in digital radiography is the:
    A. EI
    B. Target EI (EI_T)
    C. DI
    D. Volume of interest (VOI)
22. A measure of the detector response to radiation in the relevant image region of an image acquired with a digital x-ray imaging system is referred to as the:
    A. EI
    B. Target EI (EIT)
    C. DI
    D. Volume of interest (VOI)

23. Which of the following DI numbers of a DR system is able to deliver the EI_T established by the department?

    A. A DI = +3
    B. A DI = −3
    C. A DI value between +1 and −1
    D. A DI value between +2 and −2

24. The dose in digital radiography is directly proportional to the:

    A. kV^3
    B. mA^2
    C. mAs × kV
    D. mAs

25. The dose in digital radiography is also directly proportional to the:

    A. Square of the ratio of the kV
    B. Square of the ratio of the mAs
    C. Distance of the patient from the x-ray tube
    D. mAs × kV

26. The American College of Radiology (ACR) defines the Diagnostic Reference Level as:

    A. An investigation level to identify unusually high radiation dose or exposure levels for common diagnostic medical x-ray procedures
    B. A reference dose level for chest x-ray examination
    C. A dose limit for the patient
    D. A radiation protection principle to keep all exposures as low as reasonably achievable

27. The following percentile point used for DRLs indicate that image quality may be compromised:

    A. 100th percentile
    B. 75th percentile
    C. 25th percentile
    D. None of the above

28. The ACR-AAPM-SPR for example, recommends a DRL for an adult PA chest (23 cm thickness) with grid to be:

    A. 0.15 mGy
    B. 3.4 mGy
    C. 4.2 mGy
    D. 5.0 mGy

Chapter 3

29. According to the International Commission on Radiological Protection (ICRP), the second part of optimization (the first part being observing the ALARA principle) must include the:

    A. Imaging equipment and adequacy of the equipment
    B. Technical Parameters
    C. DRLs
    D. All of the above

30. The following are optimization strategies used in digital radiography except:

    A. Optimization of the exposure technique factors
    B. Optimization of the EI and DI
    C. Optimization of image processing
    D. Patient shielding

31. The major conclusion of research on the optimization of kV in digital radiography showed that:

    A. kV does not influence the magnitude of the dose to the patient
    B. The patient dose cannot be controlled using kV
    C. The kV can be reduced without compromising the image quality
    D. The optimum kV for all digital radiography examinations is 120 kV

32. An optimization of the EI in digital radiography study by Seeram et al. [16], essentially concluded that:

    A. Operators can optimize the dose to the patient by manipulating the mAs and associated EIs
    B. The EI is the patient dose
    C. An EI value of 300 indicates that the image quality is poor
    D. All of the above

33. Research examining the use of clinical EI, Target EI (EI_T) and DI as tools for optimizing the patient dose concluded that:

    A. Optimization is possible only if the EI_T is periodically and properly updated by the department
    B. EI_T is not related to image quality optimization required by departments
    C. The kV and mAs are the only factors influencing optimization of dose in D = digital radiography
    D. The DI cannot be used as a dose reduction strategy

34. Research on the use of image processing as an optimization strategy in digital radiography generally show that:

    A. Preprocessing is can be used to optimize the dose
    B. Multifrequency processing can allow a low-dose image with poor image quality (e.g., high noise) to be processed using specialized software to improve visibility of structures

C. Optimizing image processing parameters can significantly reduce the dose without compromising image quality
D. B and C are correct

35. Which of the following is a useful tool for optimization of radiation protection to identify high patient doses for various common examinations using specific imaging equipment?

A. The DI and the EI
B. The EI_T
C. DRLs
D. All of the above

Chapter 4

36. The first clinically useful CT scanner was invented by:

A. Allan Cormack
B. Godfrey Hounsfield
C. Siemens Medical Systems
D. Lambert and Beer

37. Radiation attenuation is defined as:

A. The calculation of CT numbers
B. The production of scattered radiation
C. A decrease in the intensity of x-rays as they pass through an object (patient).
D. An increase in the intensity of x-rays as they pass through an object

$$I = I_0 e^{-\mu x}$$

38. The algebraic expression shown above is important to CT, and is know as:

A. Eulers law
B. Hounsfield's law
C. Lambert-Beer's law
D. Law of x-ray intensities

39. The CT number is:

A. A number calculated for each voxel (volume element) in the slice of tissue to be imaged based on the attenuation coefficients of the tissues within the voxel.
B. A number that denotes how many attenuation coefficients are in the tissue voxel
C. A number indicating how many electrons are in the tissue voxel
D. A number used by the pioneer Hounsfield to describe the degree of radiation transmission through the patient

40. Which of the following algebraic expressions is used to calculate CT numbers, where the attenuation coefficients for tissue is μ_t; and the attenuation coefficient for water is μ_w? The manufacturer's contrast factor is K
    A. $CT = \mu_w - \mu_t/\mu_w \times K$
    B. $CT = \mu_t - \mu_w/\mu_{w.} \times K$
    C. $CT = \mu_t - \mu_w/\mu_w \times K.$
    D. $CT = \mu_w + \mu_t \times K$

41. The primary purpose of a bow-tie filter in CT is to:
    A. Shape the x-ray beam from the tube to make it more uniform at the detector
    B. Filter the x-ray beam to reduce the dose to the patient
    C. The ensure that the x-ray photons are filtered to get rid of high energy photons
    D. All of the above

42. Which of the following algorithms uses a convolution filter (digital filter) to solve the major problem of the first CT reconstruction algorithm?
    A. Back projection algorithms
    B. Filtered back projection (FBP) algorithms.
    C. Iterative reconstruction (IR) algorithms
    D. Artificial Intelligence (AI) based algorithms

43. Iterative algorithms were developed to:
    A. Reduce the noise from low dose CT imaging
    B. Allow technologists to scan the patient very quickly
    C. Minimize radiation dose to patients
    D. A and C are correct.

44. AI-based CT image reconstruction algorithms are intended to:
    A. Overcome the problems (image noise) of the FBP algorithm in low-dose CT imaging
    B. Overcome the problems of noise texture of IR algorithms in low-dose CT imaging
    C. Improve image quality, dose performance, and reconstruction speed compared with iterative reconstruction algorithms
    D. All of the above

45. A generalized framework for an AI-based such as Deep Learning Image Reconstruction (DLIR) algorithm include:
    A. Development of the algorithm
    B. Train and optimize the algorithm
    C. Performance verification of the algorithm
    D. All of the above

46. Multislice CT (MSCT) data acquisition involves:

    A. Stop-and-go data acquisition
    B. Slice-by-slice data acquisition
    C. Move the table to scan a volume of tissue using a 1Dimensional (1D) detector array that rotates at the same time the table moves through the gantry aperture
    D. Move the table while the x-ray tube and detectors rotate simultaneously around the patient using 2D detector array to collect multiple slices per revolution.

47. Which of the following CT detectors convert x-ray photons directly into electron hole pairs (electric charge)?

    A. Energy-integrating detectors
    B. Photon counting detectors
    C. Scintillation detectors
    D. Dual layer detectors

48. The following are *selectable scan parameters* except:

    A. Scan mode, and exposure factors (kV, mA, and scan time),
    B. Gantry rotation time,
    C. Pitch, scan length, collimation, and slice width.
    D. Placement of the body part in the isocenter of the gantry aperture

49. The International Electrotechnical Commission (IEC) defines the pitch in a multi-slice scanner as:
    A. P = distance the table travels per rotation (d)/total collimation (W).
    B. P = d + W
    C. P = d × W
    D. P = W/d

Chapter 5

50. Which of the following is the current international standard for estimating CT radiation dose?

    A. Thermo-luminescent dosimetry (TLD)
    B. Computed Tomography Dose Index (CTDI) dosimetry
    C. Pencil ionization chamber dosimetry
    D. Dose-Area Product dosimetry

51. The $CTDI_{volume}$ ($CTDI_{vol}$) can be calculated using the following algebraic expression:

    A. $CTDI_{vol}$ = Weighted CTDI ($CTDI_w$)/pitch
    B. Pitch × $CTDI_w$
    C. Pitch × $CTDI_{vol}$
    D. $CTDI_{vol}$ × $CTDI_w$ × Pitch

52. In CT, the dose length product (DLP) is calculated using the following algebraic expression:

    A. DLP = Pitch x Scan Length (L)
    B. DLP = Pitch x $CTDI_{vol}$
    C. DLP = $CTDI_{vol} \times L$
    D. DLP = $CTDI_{w} \times$ Pitch

53. The quantity used in CT dose reporting to relate radiation exposure to risk is:

    A. Effective Dose (ED)
    B. Absorbed Dose
    C. Entrance Surface Exposure (ESE)
    D. CTDI

54. Which of the following image quality factors in CT addresses the random variation of CT numbers in the image?

    A. Spatial resolution
    B. Contrast-to-Noise resolution
    C. Noise
    D. Artifacts

55. Noise in CT is:

    A. Inversely proportional to square root of radiation dose
    B. Inversely proportional to tube voltage
    C. Inversely proportional to fourth power of spatial resolution
    D. All of the above

56. In CT, the ability to distinguish between different CT numbers is defined as the:

    A. Contrast-to-noise resolution
    B. Noise
    C. Spatial resolution
    D. Artifacts

57. Spatial resolution in CT is defined as:

    A. The ability to distinguish small details of object
    B. The ability to distinguish between different CT numbers.
    C. A random variation of CT numbers
    D. Directly proportional to focal spot size and detector collimation

58. The key relationships that are noteworthy in CT dose optimization are:

    A. Dose is directly proportional to the mAs, that is, doubling the mAs doubles the dose
    B. The dose is proportional to the square of the kilovolt peak (kV^2).
    C. Dose is inversely related to the pitch (P).
    D. All of the above

59. Which of the following is an effort to reduce the dose to the patient at the beginning and end of the scanning?

    A. Overbeaming
    B. Overranging
    C. Adaptive or dynamic collimation
    D. Additional bow tie filtration of the primary beam from the x-ray tube

Chapter 6

60. Objective physical measures in dose optimization studies are used to assess

    A. Modulation transfer function (MTF) of the imaging system
    B. Signal-to-Noise Ratio (SNR), the Noise power spectrum (Wiener spectrum)
    C. Detective Quantum Efficiency (DQE).
    D. All of the above

61. In a dose optimization study, which approach is best suited to assess clinical image quality?

    A. Objective physical measures
    B. Observer performance methods
    C. Subjective impression of image quality parameters seen on clinical images
    D. B and C of the above

62. Which of the following approaches is best suited for lesion detection in optimization research?

    A. Visual Grading Analysis (VGA)
    B. Receiver Operating Characteristics (ROC)
    C. Increased intensity of the viewing monitor
    D. Use of a smoothing digital filter

63. Which of the following approaches is best suited for visualization and reproduction of anatomical structures in optimization research?

    A. Visual Grading Analysis (VGA)
    B. Receiver Operating Characteristics (ROC)
    C. Increased intensity of the viewing monitor
    D. Use of a smoothing digital filter

Answers to Review Questions

1. C
2. A
3. A
4. D
5. A
6. C
7. A
8. B

9. D
10. D
11. A
12. C
13. B
14. C
15. D
16. D
17. C
18. A
19. C
20. B
21. C
22. A
23. C
24. D
25. A
26. A
27. C
28. A
29. D
30. D
31. C
32. A
33. A
34. D
35. C
36. B
37. C
38. C
39. A
40. C
41. A
42. B
43. D
44. D
45. D
46. D
47. B
48. D
49. A
50. B
51. A
52. C
53. A

54. C
55. D
56. A
57. A
58. D
59. C
60. D
61. B
62. B
63. A

Index

A

Attenuation, 41–45, 93, 94

B

Biological effects of radiation exposure, 1–2

C

Cancer risks from CT imaging, 59, 60
Computed radiography (CR), 5, 13, 15–17, 19–21, 26–29, 84, 85, 88, 89
Computed tomography (CT), 4–9, 37, 41–54, 57–72, 78–81, 85, 88, 93–96
CT dose optimization, 50, 61, 63, 65, 71, 72, 96
CT dose reduction, 65, 71–72
CT numbers, 43–44, 63, 79, 80, 93, 94, 96

D

Data acquisition, 5, 16, 41, 42, 44–46, 50, 51, 95
Deep learning image reconstruction algorithms in CT, 48, 49
Deep learning image reconstruction (DLIR), 48–50, 68, 94
Detector response to radiation, 21, 90
Deterministic effects, 2, 3, 6, 58, 61, 88
Deviation index (DI), 9, 15, 20–22, 25, 30–31, 37, 88–93
Diagnostic reference level (DRL), 4, 22–23, 25, 31, 36–37, 61, 66–67, 88, 91–93
Digital radiography (DR), 4–6, 9, 13–23, 25–37, 41, 58, 65, 84–85, 88–92
Direct conversion digital radiography, 18–19
Dose and image quality optimization, 71
Dose creep in digital radiography, 6–8
Dose/image quality optimization, 77–85
Dose in digital radiography, 22, 30, 91
Dose optimization, 4, 6, 19, 22, 23, 30, 31, 33–37, 41, 42, 47, 53, 54, 58–61, 63, 65–66, 72, 79, 84–85, 97
Dose optimization in digital radiography, 4, 6–8, 22, 35, 85
Doses in computed tomography (CT), 6–9, 41, 46, 47, 53, 57–72, 78, 80

E

Exposure indicator (EI), 6, 9, 15, 20–22, 25–31, 37, 84, 85, 88–90, 92, 93
Exposure technique factors, 4, 9, 15, 25–28, 37, 63, 84, 85, 89, 92

F

Flat-panel digital radiography (FPDR), 5, 13, 15, 17–21, 26, 89, 90

I

Image display, 5, 13, 16, 19, 42, 49, 90
Image post processing, 25, 35, 42
Image processing, 6, 9, 13, 14, 16, 17, 19–20, 22, 31–33, 44, 46, 82, 89, 92, 93

E. Seeram, *Dose Optimization in Digital Radiography and Computed Tomography*,
https://doi.org/10.1007/978-3-031-22871-1

Image quality assessment tool, 33, 80–81
Image quality descriptors, 79–80
Image reconstruction algorithms, 46, 48–49, 51, 63, 68, 69, 94
Image reconstruction optimization, 47
Image storage, 42, 49
Indirect conversion digital radiography, 17–18
International Commission on Radiological Protection (ICRP), 2–4, 13, 22, 23, 25, 36, 61, 65–67, 85, 87, 92
Iterative reconstruction algorithms, 46–47, 49, 65, 70, 94

J
Justification, 3, 87, 88

K
Kilovolt (kV), 6, 64, 96
kV optimization, 26, 27, 37, 92

M
Milliampere-seconds (mAs), 6
Multifrequency processing, 32–35, 37, 92
Multislice CT detectors, 41, 64, 95
Multislice CT principles, 50–53

N
Noise reduction algorithms, 9, 31–35, 37

O
Observer, 13, 19, 29, 32, 49, 65, 77–83, 85, 89
Observer performance method, 65, 79, 81, 97
Optimization, 1–9, 22, 25–37, 48, 49, 65–72, 78, 80, 82, 84, 87, 88, 92, 93, 97
Optimization strategies, 4, 9, 19, 25–37, 57–72, 92

P
Photon counting detectors (PCD), 51–52, 66, 70–71, 95
Photostimulable luminescence, 16, 89
Physical principles, 5, 15–19, 50
Pitch, 9, 53, 54, 62–65, 95, 96
Principles of radiation protection, 2–6

R
Radiation protection, 1–9, 22, 25, 36, 37, 41, 53–54, 59, 61, 65–67, 71, 78, 85, 88, 91, 93
Radiation risks, 1, 6, 8, 58, 59
Receiver operating characteristic (ROC), 79, 81–83, 97
Research, 1, 4, 9, 22, 26–28, 30, 33, 57, 58, 71, 72, 77–85, 92, 97

S
Scientific method, 77, 84
Selectable scan parameters, 50, 53, 95
Standardized exposure indicator, 19–21
Stochastic effects, 2, 3, 6, 58, 61, 62, 88

T
Tube current modulation, 65

V
Visual grading analysis (VGA), 33, 65, 79, 81–83, 97